Energy and Agriculture: Science, Environment, and Solutions
Lab Manual

Energy and Agriculture: Science, Environment, and Solutions Lab Manual

Stephen D. Butz

Australia • Brazil • Mexico • Singapore • United Kingdom • United States

Energy and Agriculture: Science, Environment, and Solutions Lab Manual
Stephen D. Butz

Vice President, Careers & Computing: Dawn Gerrain

Director of Learning Solutions: Steve Helba

Product Director: Erin Brennan

Product Manager: Nicole Sgueglia

Director, Development-Career and Computing: Marah Bellegarde

Managing Editor: Juliet Steiner

Senior Content Developer: Darcy M. Scelsi

Editorial Assistant: Scott Royael

Brand Manager: Kay Stefanski

Senior Production Director: Wendy Troeger

Production Manager: Mark Bernard

Content Project Manager: Brooke Greenhouse

Senior Art Director: David Arsenault

Cover image(s): To come

Library of Congress Control Number: 2013950837

ISBN-13: 978-1-111-54110-1

Delmar
5 Maxwell Drive
Clifton Park, NY 12065-2919
USA

Cengage Learning is a leading provider of customized learning solutions with office locations around the globe, including Singapore, the United Kingdom, Australia, Mexico, Brazil, and Japan. Locate your local office at: **international.cengage.com/region**

Cengage Learning products are represented in Canada by Nelson Education, Ltd.

To learn more about Delmar, visit **www.cengage.com/delmar**

Purchase any of our products at your local college store or at our preferred online store **www.cengagebrain.com**

Notice to the Reader
Publisher does not warrant or guarantee any of the products described herein or perform any independent analysis in connection with any of the product information contained herein. Publisher does not assume, and expressly disclaims, any obligation to obtain and include information other than that provided to it by the manufacturer. The reader is expressly warned to consider and adopt all safety precautions that might be indicated by the activities described herein and to avoid all potential hazards. By following the instructions contained herein, the reader willingly assumes all risks in connection with such instructions. The publisher makes no representations or warranties of any kind, including but not limited to, the warranties of fitness for particular purpose or merchantability, nor are any such representations implied with respect to the material set forth herein, and the publisher takes no responsibility with respect to such material. The publisher shall not be liable for any special, consequential, or exemplary damages resulting, in whole or part, from the readers' use of, or reliance upon, this material.

Printed in the United States of America
1 2 3 4 5 6 7 18 17 16 15 14

Table of Contents

Preface

The *Energy and Agriculture Lab Manual* is a practical, hands-on lab manual designed to complement the topics presented in the *Energy and Agriculture* textbook. The 27 activities in this lab manual are practical, easy to administer, and time-tested in the classroom. They provide new, innovative ways to teach the subjects presented in the textbook. Many of the labs presented in the *Energy and Agriculture Lab Manual* are designed to be extremely flexible, and require easily obtained materials that are available to all educators. Any educator who needs to teach students about energy production through hands-on investigation will find this lab manual extremely useful.

Lab Safety Guidelines

1. Follow all instructions clearly.

2. Identify the location of all safety equipment, and be familiar with the procedures for using it in the laboratory or classroom.

3. Never eat or drink while performing laboratory experiments.

4. Report all injuries or accidents to your instructor immediately.

5. No running or horseplay in the lab.

6. Do not perform unauthorized experiments and use equipment only as directed.

7. Return equipment and supplies where you got them, and dispose of materials as instructed.

8. Keep your work area clean and uncluttered.

9. Wear appropriate clothing while performing a lab. Beware of loose sleeves, long hair, and open-toed shoes.

10. Wear safety glasses at all times when instructed.

11. Be careful with hot materials or chemicals of any kind.

12. Be careful when using fire or a heat source of any kind.

13. Allow materials and container to cool thoroughly before handling them.

14. Never smell chemicals or breathe in fumes of any kind.

15. Report spills or breaks to your instructor immediately.

16. Handle all specimens with care.

17. Never throw anything.

18. Never taste any chemical or substance of any kind.

19. Use care when using sharp or pointed objects.

20. Never pick up any broken glass with your hands.

21. Never operate electrical equipment near water or with wet hands.

22. Wash your hands thoroughly after performing any lab activity.

LAB 1
Origins of Agricultural Crops

Purpose

The purpose of this lab is to identify the regions of the world where major agricultural crops have originated. Many crops that have been used traditionally for food or forage also can be used as source of bioenergy. Identifying the source regions of crops may help to identify wild varieties of crop species that could provide beneficial traits to improve insect and disease resistance.

Materials

Scissors
Glue sticks
Colored pencils

Procedure

1. Using your colored pencils, color each of the crop types shown in Figure 1-1 a different color (Cereal/Grain/Legume, Root/Tuber/Vegetable, Forage, Tree, Sugar, and Oil).

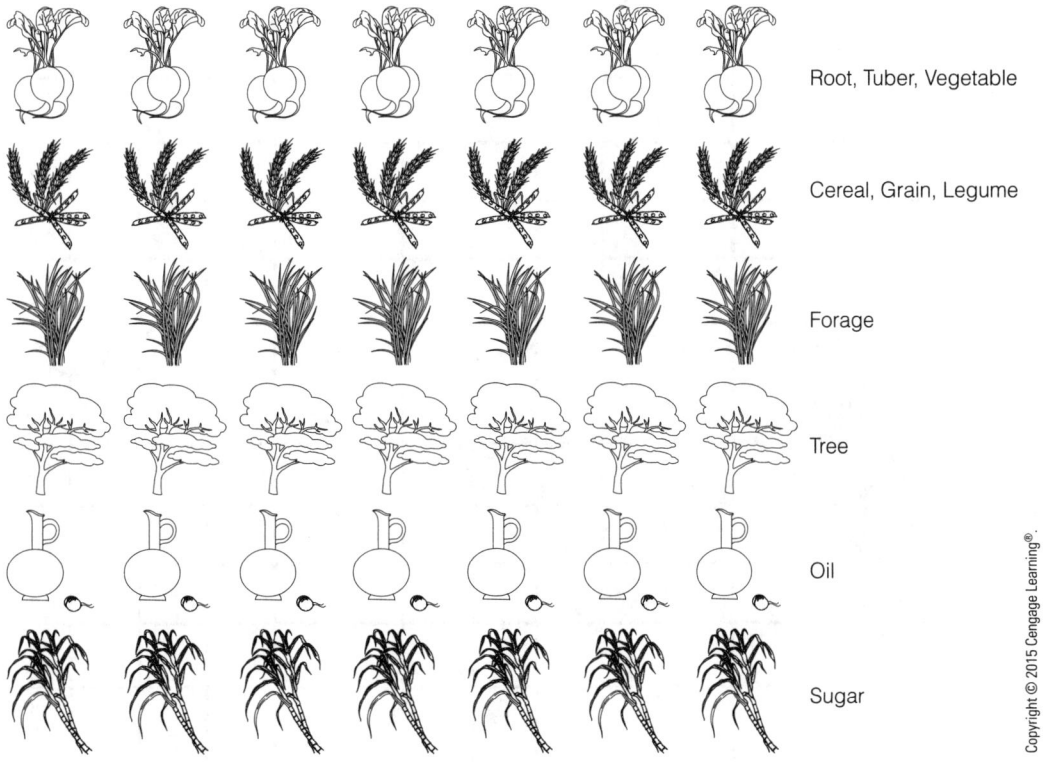

Root, Tuber, Vegetable

Cereal, Grain, Legume

Forage

Tree

Oil

Sugar

FIGURE 1-1

2. Using the latitude and longitude coordinates shown in Table 1-1, cut out and paste each of the crops in Figure 1-1 to its correct geographic place of origin on the map in Figure 1-2.

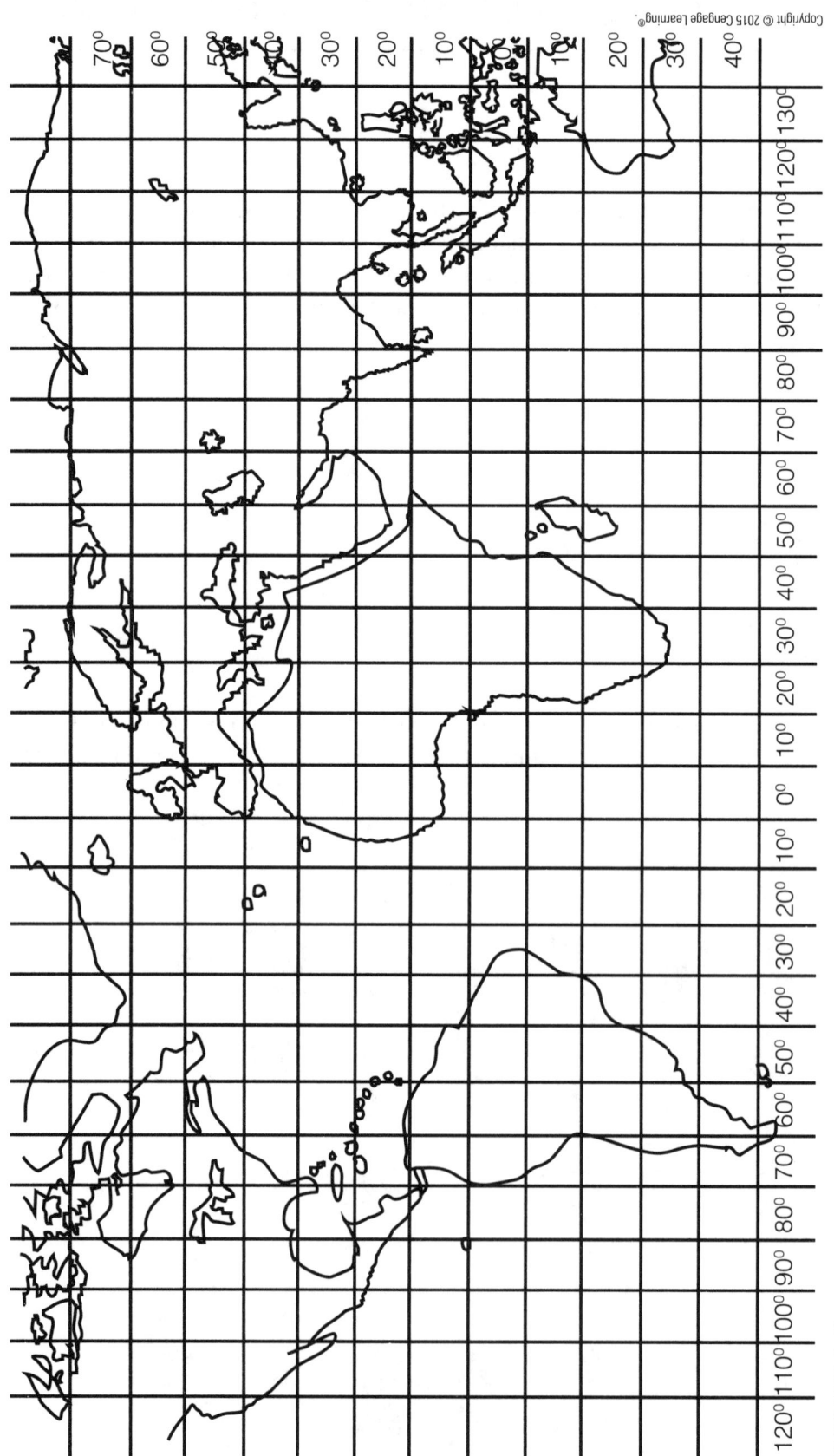

FIGURE 1-2

	Crop	Crop Classification	Latitude	Longitude
	TABLE 1-1 Latitude and Longitude Coordinates			
1	Beet	Root/Tuber/Vegetable	44° N	5° E
2	Millet	Cereal/Grain/Legume	30° N	30° E
3	Sorghum	Cereal/Grain/Legume	22° N	33° E
4	Switchgrass	Forage	40° N	100° W
5	Shrub Willow	Tree	40° N	80° W
6	Miscanthus	Forage	20° N	30° E
7	Sunflower	Oil	25° N	105° W
8	Poplar	Tree	40° N	90° W
9	Corn	Cereal/Grain/Legume	20° N	100° W
10	Peanut	Cereal/Grain/Legume	0°	75° W
11	Oil Palm	Oil	10° S	15° E
12	Sugarcane	Sugar	25° N	115° E
13	Soybean	Cereal/Grain/Legume	35° N	110° E
14	Camphor Laurel	Oil	15° N	105° E
15	Safflower	Oil	10° N	80° E
16	Sugar Beet	Sugar	43° N	15° E
17	Coconut Palm	Oil	5° N	100° E
18	Canary Grass	Forage	40° N	25° E
19	Mustard	Oil	40° N	0°
20	Rapeseed (Canola)	Oil	37° N	30° E
21	Olive	Oil	38° N	18° E
22	Wheat	Cereal/Grain/Legume	30° N	45° E

Conclusions

1. Which seven crops do you think would be best grown in cooler or temperate climates? Why?

2. Using your answer from question 1, calculate what percentage of all the bioenergy crops that are considered temperate.

3. Which six crops can be grown in tropical climates? How can you tell?

4. Using your answer from question 2, calculate what percentage of all the bioenergy crops are considered tropical.

5. Which crop type (Cereal/Grain/Legume, Root/Tuber/Vegetable, Forage, Tree, Sugar, and Oil) is most widely grown around the world as a source of biofuel?

6. Which six bioenergy crops would most likely not be considered food for humans?

LAB 2
Properties of Coal

Purpose

The purpose of this lab is to identify the physical properties of coal in order to evaluate their use as an energy source. Even though coal is not formed from minerals, the use of mineral-identifying properties can be useful for observing its physical properties.

Materials

1 Hand sample of the following for each group:
  Peat
  Lignite coal
  Bituminous coal
  Anthracite coal
Graduated cylinder (large enough to hold coal sample)
Streak plate
Glass plate for hardness test
Scale or balance
Magnifying glass
Colored pencils

Procedure

Part 1 – Physical Characteristics of Coal

1. Obtain one of each of the four coal samples.

2. Identify the physical characteristics for each coal sample and record in Table 2-1. Color represents the surface color of the sample. Hardness is a relative characteristic based on Moh's scale of mineral hardness using a value between 1–10. If you can scratch the sample with your fingernail, then the hardness is between 1–2. If you cannot scratch it with a fingernail, then try to use the sample to scratch the glass plate. If it can scratch glass, its hardness is 6–10. If it cannot scratch glass or be scratched by a fingernail, its hardness is 3–5.

3. Use the streak plate to determine the color of the sample when rubbed against the porcelain tile and record in Table 2-1.

TABLE 2-1 Physical Characteristics of Coal					
Type of Coal	**Color**	**Hardness**	**Streak**	**Luster**	**Layers**
Peat					
Lignite					
Bituminous					
Anthracite					

4. Luster is the way in which light reflects off the sample. If it appears shiny, then it is characterized as glassy. If it is not shiny, then it would be classified as dull.

5. Finally, inspect the samples to see if there are any layers visible

Part 2 – Density of Coal

1. Next, you will determine the density of each sample of coal.

2. Use a scale or balance to determine the mass of each sample to the nearest tenth of a gram and record it in Table 2-2.

3. Now use the displacement method to determine the volume of each sample. Add water to a known volume in a graduated cylinder and record it under "Start Volume" in Table 2-2.

4. Carefully slide the first coal sample into the cylinder and record the new volume of liquid within the graduated cylinder in Table 2-2. When determining the volume of peat, you may have to hold the sample under the surface of the water with the point of an unfolded paper clip, because it might float.

TABLE 2-2 Density of Coal					
Coal Type	**Mass (grams)**	**Start Volume of Water in Cylinder (ml)**	**End Volume of Water in Cylinder (ml)**	**Volume (ml)**	**Density of Sample (g/cm³)**
Peat					
Lignite					
Bituminous					
Anthracite					

5. Once you have determined the mass and volume of each sample, dry off the coal with a paper towel and calculate their density in grams/cubic centimeter.

6. Create a bar graph to display the results of your density determination for the four coal types.

Part 3 – Coal Reserves of the United States

Use your colored pencils to create a false color map of the main coal types that make up the coal reserves in the United States. Assign a different color for each coal type, and color in the map in Figure 2-1.

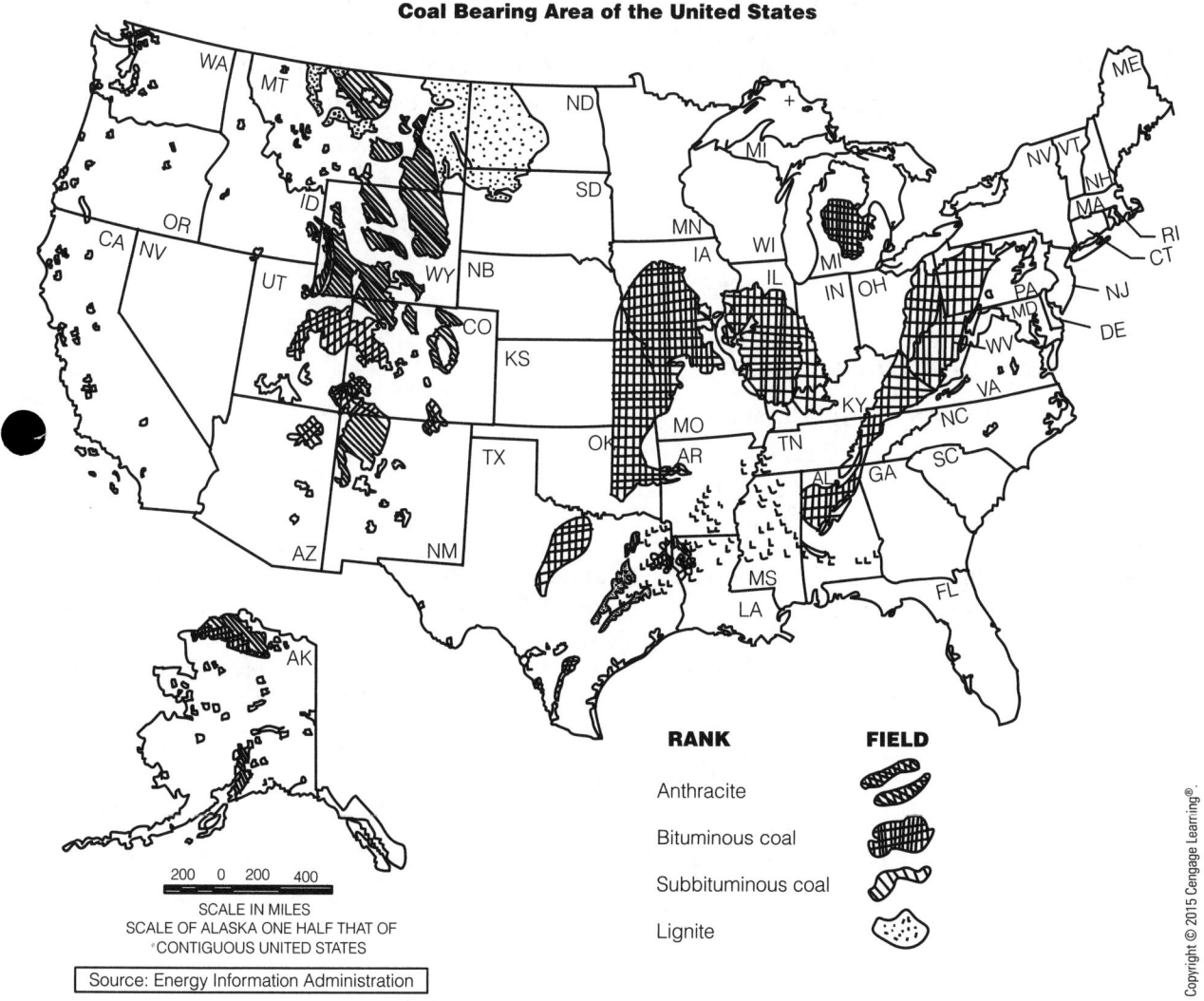

Coal Bearing Area of the United States

RANK	FIELD
Anthracite	
Bituminous coal	
Subbituminous coal	
Lignite	

200   0   200   400
SCALE IN MILES
SCALE OF ALASKA ONE HALF THAT OF
CONTIGUOUS UNITED STATES

Source: Energy Information Administration

FIGURE 2-1

Conclusions

1. Which coal type had the lowest density? The highest?

2. Did the lowest density coal also have the lowest hardness?

3. The heat content of a fuel is the amount of energy it will release when burned. Heat content is usually expressed in British Thermal Units (BTU) or kilojoules (kJ). Use the information below of the average heat content of coal to determine the heat content of each type of coal in BTUs per ton and kilojoules per ton assuming that 1 BTU = 1.055 joules.

    Average Heat Content of Coal
    Peat – 5,600 BTU/pound
    Lignite – 6,150 BTU/pound
    Bituminous – 13,000 BTU/pound
    Anthracite – 14,000 BTU/pound

4. Using your map from Part 3, which type of coal does the United States have most in reserves? Least in reserves?

5. In which states is coal found?

LAB 3
Coal Syngas

Purpose

The purpose of this lab is to observe how coal can be used to produce a combustible gas, which can be used directly or converted into liquid fuels. The gas produced is known as *syngas*, which is a mixture of hydrogen and carbon monoxide. The process to produce it is known as *destructive distillation* or *pyrolysis*, which is the exposure of organic material to high temperatures in the absence of oxygen.

Materials

Crushed coal samples
Bunsen burner or alcohol burner
Vent hood
Ring stand with test tube clamp
Pyrex test tube
Rubber stopper with hole
Glass tubing (approximately 2 inches in length)
Matches
Safety glasses

SAFETY PROCEDURES
THIS LAB WILL NEED TO BE PERFORMED IN A VENTILATION HOOD TO ALLOW FOR THE SAFE REMOVAL OF FUMES PRODUCED DURING THE EXPERIMENT. ALSO, THE USE OF SAFETY GLASSES BY EVERY STUDENT DURING THE EXPERIMENT IS REQUIRED. BECAUSE STUDENTS ARE WORKING WITH AN OPEN FLAME, MAKE SURE THEY ARE GIVEN THE PROPER SAFETY PROCEDURES FOR LIGHTING AND EXTINGUISHING A BUNSEN BURNER OR ALCOHOL BURNER. ALSO, MAKE SURE STUDENTS WITH LONG HAIR HAVE IT TIED BACK TO AVOID CONTACT WITH AN OPEN FLAME AND BEWARE OF ANY LOOSE CLOTHING THAT MAY COME IN CONTACT WITH THE FLAME. IF YOU ARE USING MATCHES, MAKE SURE TO DIP THEM IN WATER OR RUN THE MATCH HEADS UNDER WATER BEFORE DISPOSING OF THEM.

Procedure

1. Fill your test tube about ¼ full with crushed coal.

2. Carefully insert the glass tubing into the hole of the rubber stopper, assuring a tight fit. Leave about 1 to 1½ inches of the glass tube sticking out from the stopper.

3. Insert the rubber stopper tightly into the test tube.

4. Attach the test tube to the clamp and ring stand. Adjust the test tube so that it sits at a 45 degree angle.

5. Carefully light the burner.

6. Adjust the height of the ring stand so that the base of the test tube is just above the flame (see Figure 3-1).

7. When the coal is first heated, water vapor and smoke will be produced and vent out through the glass tube. After a few minutes, all the air inside the test tube will be driven out, and pyrolysis will begin.

8. Use a match to try and ignite the vapor coming out of the glass tube. It may take a few tries, but eventually it will light. Once the vapor is combustible, syngas is being produced.

9. Carefully turn off the burner, and observe the contents of the test tube. Wait until the apparatus is completely cooled before disassembling it per your teacher's instructions.

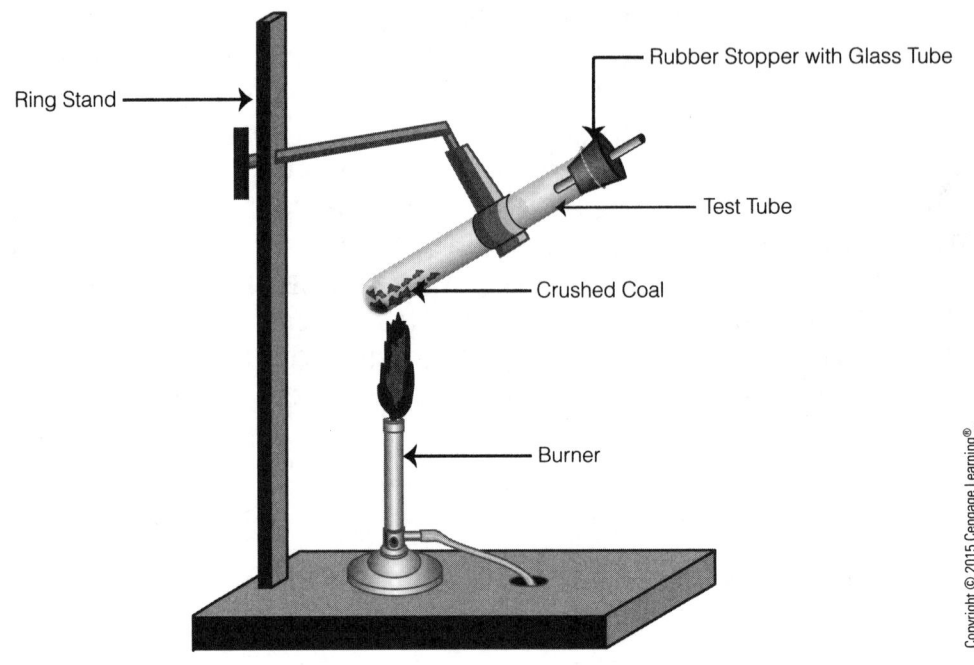

Ring Stand

Rubber Stopper with Glass Tube

Test Tube

Crushed Coal

Burner

FIGURE 3-1

Conclusions

1. Did the contents of the test tube change after pyrolysis? If so how?

2. How does pyrolysis differ from combustion?

3. What are three products produced by the destructive distillation of coal?

4. Besides being flammable, what is another danger of syngas?

LAB 4
Energy Use in the United States and the World

Purpose

The purpose of this lab is to observe how the main sources of energy production have changed in the United States beginning in 1755, and how energy use around the world varies by region and population.

Materials

Graph paper or spreadsheet software
Colored pencils

Procedure

Part 1: Historical Energy Consumption in the United States

1. Using the data from Table 4-1, calculate the total energy consumption for each year.

2. Next, use the data from Table 4-1 on the consumption of energy in the United States by source to create a multi-line graph showing the trends in energy resource use from 1775–2010. Also plot a line for the total amount of energy used for each year. Label the X-axis of your graph **Year**, and the Y-axis, **Quads**.

3. When your graph is complete, highlight each line for each of the energy resources using a different color, and make a key for your graph.

TABLE 4-1   U.S. Historical Energy Production

U.S. Energy Consumption 1775–2010
(Quadrillion BTU)

Year	Coal	Natural Gas	Petroleum	Hydropower	Wood	Nuclear	Total
1775	0.0	0.0	0.0	0.1	0.2	0.0	
1785	0.0	0.0	0.0	0.1	0.3	0.0	
1795	0.0	0.0	0.0	0.1	0.4	0.0	
1805	0.0	0.0	0.0	0.1	0.5	0.0	
1815	0.0	0.0	0.0	0.1	0.7	0.0	
1825	0.0	0.0	0.0	0.1	1.0	0.0	
1835	0.0	0.0	0.0	0.1	1.3	0.0	
1845	0.0	0.0	0.0	0.1	1.8	0.0	
1855	0.4	0.0	0.0	0.1	2.4	0.0	
1865	0.6	0.0	0.0	0.1	2.8	0.0	
1875	1.4	0.0	0.0	0.1	2.9	0.0	
1885	2.8	0.1	0.0	0.1	2.7	0.0	
1895	5.0	0.1	0.2	0.1	2.3	0.0	
1905	10.0	0.4	0.6	0.4	1.8	0.0	
1915	13.3	0.7	1.4	0.7	1.7	0.0	
1925	14.7	1.2	4.3	0.7	1.5	0.0	
1935	10.6	1.9	5.7	0.8	1.4	0.0	
1945	16.0	3.9	10.1	1.4	1.3	0.0	
1955	11.2	9.0	17.3	1.4	1.4	0.0	
1965	11.6	15.8	23.2	2.1	1.3	0.0	
1975	12.7	19.9	32.7	3.2	1.5	1.9	
1985	17.5	17.7	30.9	3.0	2.7	4.1	
1995	20.1	22.7	34.4	3.2	2.4	7.1	
2005	22.8	22.6	40.4	2.7	2.1	8.2	
2010	19.8	23.4	35.3	2.7	1.9	8.3	

Part 2: World Energy Use and Population

1. Use the data from Table 4-2 to create a dual axes, create a line graph showing the trends in world energy use and population.

2. The X-axis of your graph should be labeled Country/Region, and the primary Y-axis labeled Percent of World Energy Use.

TABLE 4-2 World Energy Use vs. Population		
Country/Region	Percent of World Energy Use (Quads)	Percent of World Population
USA	18.4	4.5
China	20.4	19.3
India	5	17.7
Russia	5.6	2
Japan	4	1.8
Canada	3	0.5
Middle East	5.4	0.1
Africa	3.8	0.1
Central and South America	5.4	0.1

3. Plot the data for world energy use, and then set up your secondary Y-axis for Percent World Population.

4. Once you have plotted both sets of data, use colored pencils to highlight each line and make a key for your graph.

Conclusions

1. What two sources of energy have been in use since 1775?

2. Which energy source increased rapidly shortly after the American Civil War?

3. What year did petroleum begin to take over coal as the main source of energy in the United States?

4. Place the order of each energy source from greatest to least for the following years: 1925, 1945, and 2010.

5. What is the general shape of the energy use curves from 1775 to 2010 (linear, exponential, bell-shaped)?

6. Why did coal use drop in the 1930s?

7. What caused coal to drop in the 1940s?

8. What percentage of total energy do fossil fuels make up out of the total consumption of energy for 2010?

9. How long did it take for total energy consumption to approximately double from 1935?

10. How long did it take for total energy consumption to approximately double from 1955?

11. Why do you think energy consumption dropped from 2005–2010?

12. During what years was coal the main source of energy for the United States? What period of time is it known as?

13. Which energy resources are currently increasing in their use?

14. During which year did oil use drop? Why did this occur?

15. Which energy resource is replacing coal and petroleum? Is this a renewable resource?

16. Calculate the percent change in the use of coal and petroleum between 1955 and 2005.

17. Which country uses the greatest amount energy per person? The least per person?

LAB 5
Carbon Sequestration

Purpose

The purpose of this lab is to observe an example of the chemical sequestration of carbon dioxide by using limewater (calcium hydroxide). You will then calculate how much carbon dioxide can be potentially sequestered by trees in a managed forest ecosystem.

Materials

100 ml Limewater (calcium hydroxide)
125 ml Erlenmeyer flask
Rubber stopper with two holes
Glass or hard plastic tubes
Plastic or rubber tubing
Small glass funnel
Ring stand test tube clamp
Vacuum aspirator or vacuum pump
Alcohol burner
Safety glasses

SAFETY PROCEDURES
THIS LAB WILL NEED TO BE PERFORMED IN A VENTILATION HOOD TO ALLOW FOR THE SAFE REMOVAL OF FUMES PRODUCED DURING THE EXPERIMENT. ALSO, THE USE OF SAFETY GLASSES BY EVERY STUDENT DURING THE EXPERIMENT IS REQUIRED. BECAUSE STUDENTS ARE WORKING WITH AN OPEN FLAME, MAKE SURE THEY ARE GIVEN THE PROPER SAFETY PROCEDURES FOR LIGHTING AND EXTINGUISHING AN ALCOHOL BURNER. ALSO, MAKE SURE STUDENTS WITH LONG HAIR HAVE IT TIED BACK TO AVOID CONTACT WITH AN OPEN FLAME.

Procedure

Part 1: Chemical Carbon Sequestration

1. Gather the required materials, and set up your apparatus with the instructions given by your teacher and by using the diagram in Figure 5-1.

2. Observe the clarity of the limewater within the flask.

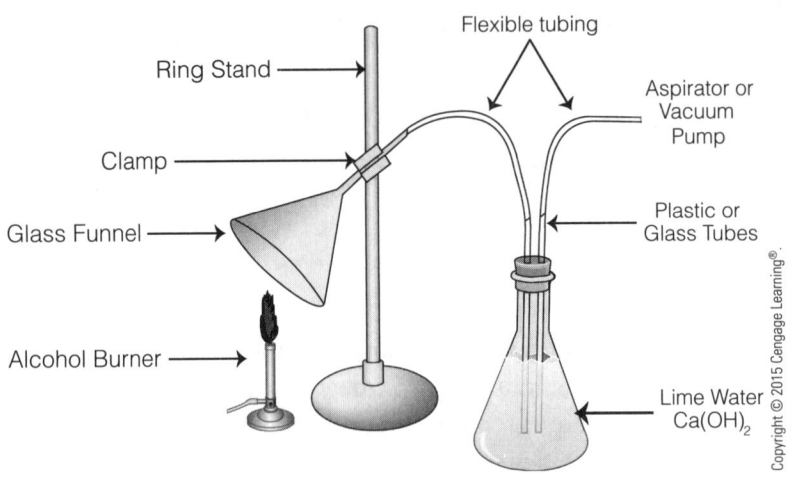

FIGURE 5-1

3. Turn on the aspirator or vacuum pump to begin drawing air through the limewater.

4. Carefully light the alcohol burner and adjust the position of the glass funnel so the fumes are drawn into it.

5. Observe the demonstration for about 5 minutes, and watch how the carbon dioxide produced from the flame affects the limewater.

6. Once the carbon sequestration has occurred, extinguish the flame according to your teacher's instructions, and disassemble your apparatus.

Part 2: Terrestrial Carbon Sequestration in Forestland

1. Use the Forest Carbon Sequestration Worksheet (Table 5-1) to determine the amount of carbon dioxide that can be sequestered by an example forest using the following information.

2. The forest of study contains the following trees: 90 Sugar Maples planted 20 years ago, 40 Red Maples planted 20 years ago, 20 Shagbark Hickories planted 15 years ago, and 20 White Oaks planted 10 years ago.

3. Using Table 5-1, fill in the following information about the Sugar Maples within the forest.

4. In column A, use Table 5-2 to fill in the name of the tree species (Sugar Maple), its Type, and Growth Rate.

TABLE 5-1 Forestry Carbon Sequestration Worksheet (U. S. Dept. of Energy, Energy Information Administration)

Reporting Year: _____

Species Characteristics (refer to Table 1)			B. Tree Age	C. Number of Age 0 Trees Planted	D. Survival Factor (Refer to Table 2)	E. Number of Surviving Trees (C × D)	F. Annual Sequestration Rate (lbs/tree) (Refer to Table 2)	G. Carbon Sequestered (lbs) (E × F)
A. Name	Tree Type (H or C)	Growth Rate (S, M, or F)						

Total Pounds of Carbon Sequestered	
Total Pounds of Equivalent CO_2 Sequestered	×3.67
Equivalent CO_2 Sequestered in Short Tons	/2,000

19

TABLE 5-2 Common Tree Species (U. S. Dept. of Energy, Energy Information Administration)

Species	Type	Growth Rate	Species	Type	Growth Rate
Ailanthus, *Ailanthus altissima*	H	F	Elm, Chinese, *Ulmus parvifolia*	H	M
Alder, European, *Alnus glutinosa*	H	F	Elm, rock, *Ulmus thomasii*	H	S
Ash, green, *Fraxinus pennsylvanica*	H	F	Elm, September, *Ulmus serotina*	H	F
Ash, mountain, American, *Sorbus americana*	H	M	Elm, Siberian, *Ulmus pumila*	H	F
Ash, white, *Fraxinus americana*	H	F	Elm, slippery, *Ulmus rubra*	H	M
Aspen, bigtooth, *Populus grandidentata*	H	M	Fir, balsam, *Abies balsamea*	C	S
Aspen, quaking, *Populus tremuloides*	H	F	Fir, Douglas, *Pseudotsuga menziesii*	C	F
Baldcypress, *Taxodium distichum*	C	F	Ginkgo, *Ginkgo biloba*	H	S
Basswood, American, *Tilia americana,*	H	F	Hackberry, *Celtis occidentalis*	H	F
Beech, American, *Fagus grandifolia*	H	S	Hawthorne, *Crataegus* spp.	H	M
Birch, paper (white), *Betula papyrifera*	H	M	Hemlock, eastern, *Tsuga canadensis*	C	M
Birch, river, *Betula nigra*	H	M	Hickory, bitternut, *Carya cordiformis*	H	S
Birch, yellow, *Betula alleghaniensis*	H	S	Hickory, mockernut, *Carya tomentosa*	H	M
Boxelder, *Acer negundo*	H	F	Hickory, shagbark, *Carya ovata*	H	S
Buckeye, Ohio, *Aesculus glabra*	H	S	Hickory, shellbark, *Carya laciniosa*	H	S
Catalpa, northern, *Catalpa speciosa*	H	F	Hickory, pignut, *Carya glabra*	H	M
Cedar-red, eastern, *Juniperus virginiana*	C	M	Holly, American, *Ilex opaca*	H	S
Cedar-white, northern, *Thuja occidentalis*	C	M	Honeylocust, *Gleditsia triacanthos*	H	F
Cherry, black, *Prunus serotina*	H	F	Hophornbeam, eastern, *Ostrya virginiana*	H	S
Cherry, pin, *Prunus pennsylvanica*	H	M	Horsechestnut, common, *Aesculus hippocastanum*	H	F
Cottonwood, eastern, *Populus deltoides*	H	M	Kentucky coffeetree, *Gymnocladus dioicus*	C	F
Crabapple, *Malus* spp.	H	M	Linden, little-leaf, *Tilia cordata*	H	F
Cucumbertree, *Magnolia acuminata*	H	F	Locust, black, *Robinia pseudoacacia*	H	F
Dogwood, flowering, *Cornus florida*	H	S	London plane tree, *Platanus_X_acerifolia*	H	F
Elm, American, *Ulmus americana*	H	F	Magnolia, southern, *Magnolia grandifolia*	H	M

(continues)

TABLE 5-2   Common Tree Species (U. S. Dept. of Energy, Energy Information Administration) (Continued)

Species	Type	Growth Rate	Species	Type	Growth Rate
Maple, bigleaf, *Acer macrophyllum*	H	S	Pine, European black, *Pinus nigra*	C	S
Maple, Norway, *Acer platanoides*	H	M	Pine, jack, *Pinus banksiana*	C	F
Maple, red, *Acer rubrum*	H	M	Pine, loblolly, *Pinus taeda*	C	F
Maple, silver, *Acer saccharinum*	H	M	Pine, longleaf, *Pinus palustris*	C	F
Maple, sugar, *Acer saccharum*	H	S	Pine, ponderosa, *Pinus ponderosa*	C	F
Mulberry, red, *Morus rubra*	H	F	Pine, red, *Pinus resinosa*	C	F
Oak, black, *Quercus velutina*	H	M	Pine, Scotch, *Pinus sylvestris*	C	S
Oak, blue, *Quercus douglasii*	H	M	Pine, shortleaf, *Pinus echinata*	C	F
Oak, bur, *Quercus macrocarpa*	H	S	Pine, slash, *Pinus elliottii*	C	F
Oak, California black, *Quercus kelloggii*	H	S	Pine, Virginia, *Pinus virginiana*	C	M
Oak, California White, *Quercus lobata*	H	M	Pine, white eastern, *Pinus strobus*	C	F
Oak, canyon live, *Quercus chrysolepsis*	H	S	Poplar, yellow, *Liriodendron tulipifera*	H	F
Oak, chestnut, *Quercus prinus*	H	S	Redbud, eastern, *Cercis canadensis*	H	M
Oak,Chinkapin, *Quercus muehlenbergii*	H	M	Sassafras, *Sassafras albidum*	H	M
Oak, Laurel, *Quercus laurifolia*	H	F	Spruce, black, *Picea mariana*	C	S
Oak, live, *Quercus virginiana*	H	F	Spruce, blue, *Picea pungens*	C	M
Oak, northern red, *Quercus rubra*	H	F	Spruce, Norway, *Picea abies*	C	M
Oak, overcup, *Quercus lyrata*	H	S	Spruce, red, *Picea rubens*	C	S
Oak, pin, *Quercus palustris*	H	F	Spruce, white, *Picea glauca*	C	M
Oak, scarlet, *Quercus coccinea*	H	F	Sugarberry, *Celtis laevigata*	H	F
Oak, swamp white, *Quercus bicolor*	H	M	Sweetgum, *Liquidambar styraciflua*	H	F
Oak, water, *Quercus nigra*	H	M	Sycamore, *Platanus occidentalis*	H	F
Oak, white, *Quercus alba*	H	S	Tamarack, *Larix laricina*	C	F
Oak, willow, *Quercus phellos*	H	M	Walnut, black, *Juglans nigra*	H	F
Pecan, *Carya illinoensis*	H	S	Willow, black, *Salix nigra*	H	F

Type: H = Hardwood, C = Conifer   Growth Rate: S = Slow, M = Moderate, F = Fast

5. Next, fill in the age of the tree in Column B, which is 20 years for the Sugar Maples.

6. In column C fill in the number of trees, which is 90 for the Sugar Maple.

7. Using the information from Column A on the Tree Type and Growth Rate, and the Survival Factors from Table 5-3, fill in the Survival Factor in column D. Because the Sugar Maple is 20 years old, has a slow growth rate, and is a hardwood, its Survival Factor is 0.448.

8. Next, calculate the Number of Surviving Trees in Column E by multiplying the number of trees by its survival factor. For the Sugar Maple the calculation will be 90×0.448.

9. Now determine the Annual Sequestration rate in pounds per tree in Column F, by using Table 5-3. For the Sugar Maples, because they are 20-year old, slow-growing hardwoods, the sequestration rate is 10.8 pounds per tree.

10. To determine the total amount of carbon sequestered by the Sugar Maples (Column G) multiply the Number of Surviving Trees (Column E) by The Annual Sequestration Rate (Column F). For the Sugar Maples the calculation would be 40.3×10.8.

11. To determine the total amount of carbon sequestered by the example forest, fill in the rest of the information for the remaining trees listed in Part 2.

12. Now add all the values in Column G to calculate the total pounds of carbon sequestered and fill in the value in Table 5-1.

13. Convert the total pounds of carbon sequestered to its equivalent carbon dioxide sequestered by multiplying the total pounds of carbon by 3.67. Record this value in Table 5-1.

14. Finally convert the value of carbon dioxide sequestered by the example forest into tons by dividing it by 2,000. Record this value in Table 5-1.

TABLE 5-3 Survival Factors and Annual Carbon Sequestration Rates for Common Trees (U. S. Dept. of Energy, Energy Information Administration)

| Tree Age (yrs) | Survival Factors by Growth Rate | | | Annual Sequestration Rates by Tree Type and Growth Rate (lbs carbon/tree/year) | | | | | |
| | | | | Hardwood | | | Conifer | | |
	Slow	Moderate	Fast	Slow	Moderate	Fast	Slow	Moderate	Fast
0	0.873	0.873	0.873	1.3	1.9	2.7	0.7	1.0	1.4
1	0.798	0.798	0.798	1.6	2.7	4.0	0.9	1.5	2.2
2	0.736	0.736	0.736	2.0	3.5	5.4	1.1	2.0	3.1
3	0.706	0.706	0.706	2.4	4.3	6.9	1.4	2.5	4.1
4	0.678	0.678	0.678	2.8	5.2	8.5	1.6	3.1	5.2
5	0.658	0.658	0.658	3.2	6.1	10.1	1.9	3.7	6.4
6	0.639	0.639	0.644	3.7	7.1	11.8	2.2	4.4	7.6
7	0.621	0.621	0.630	4.1	8.1	13.6	2.5	5.1	8.9
8	0.603	0.603	0.616	4.6	9.1	15.5	2.8	5.8	10.2
9	0.585	0.589	0.602	5.0	10.2	17.4	3.1	6.6	11.7
10	0.568	0.576	0.589	5.5	11.2	19.3	3.5	7.4	13.2
11	0.552	0.564	0.576	6.0	12.3	21.3	3.8	8.2	14.7
12	0.536	0.551	0.563	6.5	13.5	23.3	4.2	9.1	16.3
13	0.524	0.539	0.551	7.0	14.6	25.4	4.6	9.9	17.9
14	0.512	0.527	0.539	7.5	15.8	27.5	4.9	10.8	19.6
15	0.501	0.516	0.527	8.1	16.9	29.7	5.3	11.8	21.4
16	0.490	0.504	0.516	8.6	18.1	31.9	5.7	12.7	23.2
17	0.479	0.493	0.505	9.1	19.4	34.1	6.1	13.7	25.0
18	0.469	0.483	0.495	9.7	20.6	36.3	6.6	14.7	26.9
19	0.459	0.472	0.484	10.2	21.9	38.6	7.0	15.7	28.8
20	0.448	0.462	0.474	10.8	23.2	41.0	7.4	16.7	30.8
21	0.439	0.452	0.464	11.4	24.4	43.3	7.9	17.8	32.8
22	0.429	0.442	0.454	12.0	25.8	45.7	8.3	18.9	34.9
23	0.419	0.433	0.445	12.5	27.1	48.1	8.8	20.0	37.0
24	0.410	0.424	0.435	13.1	28.4	50.6	9.2	21.1	39.1
25	0.401	0.415	0.426	13.7	29.8	53.1	9.7	22.2	41.3
26	0.392	0.406	0.417	14.3	31.2	55.6	10.2	23.4	43.5
27	0.384	0.398	0.409	15.0	32.5	58.1	10.7	24.6	45.7
28	0.375	0.389	0.400	15.6	33.9	60.7	11.2	25.8	48.0
29	0.367	0.381	0.392	16.2	35.3	63.3	11.7	27.0	50.3
30	0.359	0.373	0.383	16.8	36.8	65.9	12.2	28.2	52.7
31	0.352	0.365	0.375	17.5	38.2	68.5	12.7	29.5	55.1
32	0.344	0.358	0.367	18.1	39.7	71.2	13.3	30.7	57.5
33	0.337	0.350	0.360	18.7	41.1	73.8	13.8	32.0	59.9
34	0.330	0.343	0.349	19.4	42.6	76.5	14.3	33.3	62.4
35	0.323	0.336	0.339	20.0	44.1	79.3	14.9	34.7	64.9

Conclusions

1. What changes in the physical appearance of the limewater occurred as carbon dioxide was drawn through it?

2. Write a balanced equation for the reaction of carbon dioxide gas with the limewater.

3. What changes in the phase of matter occurs when carbon dioxide is mixed with limewater?

4. How do you think the reaction of carbon dioxide with limewater could be used to sequester carbon?

5. What are some of the ways in which the calcium carbonate produced by chemical sequestration can be used?

6. How much total carbon dioxide did you determine to be sequestered by the example forest in Part 2?

7. If a school building produces 2.2 million tons of carbon dioxide each year, how many acres of 10-year-old Quacking Aspen trees would you need to sequester all the carbon dioxide produced by the school? (Assume that there are 400 Aspen trees growing per acre.)

LAB 6
Petroleum from Tar Sands

Purpose

The purpose of this lab is to learn the basic procedure for extracting petroleum from tar sand deposits. Tar sands are a mixture of bitumen and sand. Bitumen is a naturally occurring, black or brown solid hydrocarbon.

Materials

Tar sand (oil sand) sample
Small plastic bowls (small, Greek-style yogurt containers work great!)
Popsicle sticks
250 ml Beaker
Hot plate
1 Molar sodium hydroxide solution (40 grams NaOH mixed with 1 liter of distilled water)
Paper towels
Electronic balance or scale
Plastic gloves
Safety goggles

SAFETY PROCEDURES
TAR SANDS ARE A MIXTURE OF STICKY BLACK BITUMEN AND SAND. THE BITUMEN HAS A STRONG ODOR AND CAN STAIN THE SKIN AND CLOTHES. STUDENTS MUST USE PLASTIC GLOVES WHEN HANDLING THE TAR SAND AND NEED TO BE CAREFUL TO NOT STAIN THEIR CLOTHING. ALTHOUGH THIS ACTIVITY DOES NOT REQUIRE A VENTILATION HOOD, PLEASE AVOID DIRECT INHALATION OF THE VAPOR FROM THE TAR SAND. ALSO, SODIUM HYDROXIDE IS A CAUSTIC CHEMICAL THAT CAN CAUSE BURNS TO THE SKIN IF IMPROPERLY HANDLED. USE PLASTIC GLOVES AND SAFETY GOGGLES WHEN HANDLING OR WORKING WITH SODIUM HYDROXIDE.

Procedure

Part 1: Tar Sand Separation – The Hot Water Method

1. Fill a 250 ml beaker with water, and heat to near boiling by using a hot plate.

TABLE 6-1	Percentage of Bitumen Recovered			
Tar Sand Separation Method	Start Mass of Tar Sand	Mass of Sand After Separation	Mass of Bitumen	Percentage of Bitumen Recovered
Hot Water Method				
Sodium Hydroxide Method				

2. While the water is heating, place a paper towel on an electric scale or balance, and zero the scale. Weigh out an approximately 50.0 gram-sample of tar sand, and record its mass to the nearest tenth of a gram in Table 6-1.

3. Place the tar sand sample into the plastic bowl, and then carefully add hot water so that at least two-thirds of the container is filled with water.

4. Mix the water and the tar sand using a popsicle stick to separate the sand from the bitumen. If the bitumen adheres to the stick, periodically wipe the stick with a paper towel. Keep mixing and removing the bitumen until much of the bitumen is separated from the sand.

5. Carefully pour off as much of the water and bitumen from the bowl as you can. Set the bowl containing the sand aside to dry overnight.

6. After the sand has completely dried, carefully pour it out of the bowl, making sure to not mix it with any remaining bitumen, and determine its mass. Record your results in Table 6-1.

7. Subtract the dry mass of the sand, from the total start mass of the tar sand sample from the start of the experiment to determine the approximate amount of bitumen removed. Record your results in Table 6-1.

Part 2: Tar Sand Separation – The Sodium Hydroxide and Hot Water Method

1. Place a paper towel on an electric scale or balance, and zero the scale. Weigh out another approximate 50.0-gram sample of tar sand, and record its mass to the nearest tenth of a gram in Table 6-1.

2. Place the tar sand sample into the plastic bowl, and carefully add hot water so that at least two-thirds of the container is filled with water.

3. Next, add about 15 drops of the sodium hydroxide solution to the hot water and tar sand mixture.

4. Mix the solution and the tar sand using a popsicle stick to separate the sand from the bitumen. As the bitumen adheres to the stick, periodically wipe of the stick with a paper towel. Keep mixing and removing the bitumen, until much of the bitumen is separated from the sand.

5. Carefully pour off as much of the water and bitumen from the bowl as you can. Set the bowl containing the sand aside to dry overnight.

6. After the sand has completely dried, carefully pour it out of the bowl, making sure to not mix it with any remaining bitumen, and determine its mass. Record your results in Table 6-1.

7. Subtract the dry mass of the sand, from the total start mass of the tar sand sample from the start of the experiment to determine the approximate amount of bitumen removed. Record your results in Table 6-1.

Conclusions

1. Which method produced a higher amount of bitumen from the tar sand?

2. If bitumen weighs approximately 8.58 pounds per gallon, how much bitumen would you need to produce 1 barrel of oil?

3. If you used only the hot water method of bitumen extraction from tar sands, how many pounds of tar sands would you need to produce one barrel of oil?

4. If you used only the sodium hydroxide method of bitumen extraction from tar sands, how many pounds of tar sands would you need to produce one barrel of oil?

5. What are some positive and negative aspects of extracting oil from tar sands?

LAB 7
Radioactive Decay of Nuclear Fission Products

Purpose

The purpose of this lab is to observe the radioactive decay process of plutonium 239, which is one of the fission products produced by nuclear power plants. The radioactive decay of fission products occurs when radioactive atoms lose their mass in the form of alpha and beta radiation, eventually arriving at a stable form. In the case of plutonium, its stable form is lead 207. This lab will also explore the time it takes for radioactive elements to breakdown, known as their *half-lives*. Nuclear fission is used to produce approximately 19% of the electricity in the United States, and the radioactive decay products that are produced during power production are considered highly dangerous and require long-term, secure storage.

Materials

Calculator
Ruler
Colored pencils

Procedure

Part 1: Radioactive Decay of Plutonium

1. Use the information below to fill out data Table 7-1 on the radioactive decay of plutonium 239.

2. When the nucleus of an atom decays, it can emit alpha or beta radiation.

3. Alpha radiation is formed when the nucleus of an atom emits 2 protons and 2 neutrons. This causes the atomic number of the atom to decrease by 2, and the atomic mass to decrease by 4.

4. Beta radiation is formed when the nucleus of an atom emits an electron, which is formed when a neutron decays into 1 electron and 1 proton. The proton stays within the nucleus, but the electron is emitted. This causes the atomic number to increase by 1, and the atomic mass to remain the same.

TABLE 7-1 Radioactive Decay of Plutonium 239			
Radioactive Isotope	**Atomic Mass**	**Atomic Number**	**Type of Radiation**
Plutonium (Pu)	239	94	Undergoes Alpha Radiation
Uranium (U)			Undergoes Alpha Radiation
Thorium (Th)			Undergoes Beta Radiation
Protactinium (Pa)			Undergoes Alpha Radiation
Actinium (Ac)			Undergoes Beta Radiation
Thorium (Th)			Undergoes Alpha Radiation
Radium (Ra)			Undergoes Alpha Radiation
Radon (Rn)			Undergoes Alpha Radiation
Polonium (Po)			Undergoes Alpha Radiation
Lead (Pb)			Undergoes Beta Radiation
Bismuth (Bi)			Undergoes Alpha Radiation
Thallium (Th)			Undergoes Beta Radiation
Lead	207	82	

5. The first step in the decay of plutonium 239 begins when its nucleus emits alpha radiation. In Table 7-1, fill out the resulting atomic mass and number for the radioactive isotope that results.

6. Continue to fill out the radioactive decay data to complete the table. If you have done the correct math, then you should arrive at lead 207, which has an atomic mass of 207 and an atomic number of 82.

Part 2: Radioisotope Decay Graph

1. Use the data on the atomic mass and number for the decay of plutonium in Table 7-1 to create a graph illustrating its decay series.

2. Now you will create a decay chain for plutonium. To do this, plot a data point for each isotope on the graph in Figure 7-1, using its atomic number and mass. Label each data point with the element's atomic symbol, and connect each point with an arrow using a colored pencil, showing how one element decays into another.

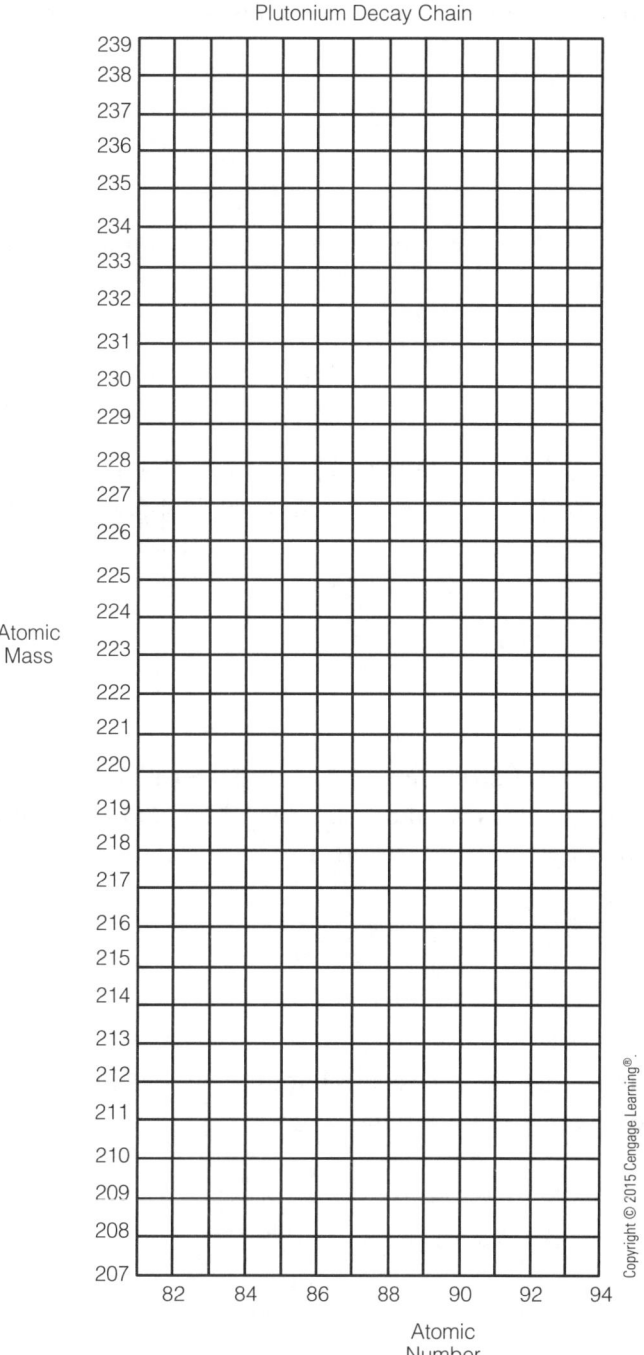

FIGURE 7-1

Conclusions

1. How many isotopes of thorium and lead exist in the decay chain of plutonium?

2. How do the different isotopes of thorium differ from each other?

3. How does alpha radiation effect the atomic mass and number of an radioactive isotope?

4. How does beta radiation effect the atomic mass and number of an radioactive isotope?

5. The half-life of plutonium is 24,000 years. This is the time it takes for half of it to decay into stable lead 207. How many years will it take 24 pounds of plutonium to decay into 6 pounds of lead?

6. Another fission product produced in nuclear power plants is cesium 137. Its half-life is 30 years. How many years would it take for 100 pounds of cesium to decay into 25 pounds of its stable form of barium 137?

7. Knowing about the decay process of plutonium and its half-life, explain why the storage of spent nuclear fuel is so important.

LAB 8
High-Temperature Solar

Purpose

The purpose of this lab is to observe how high-temperature solar power can be used as a source of energy. High-temperature solar utilizes mirrors and lenses to concentrate the Sun's energy, which can be used to generate extremely high temperatures. This heat energy can then be used to generate electricity or for industrial processes. This lab must be performed outside, on a clear, sunny day.

Materials

Thermometer
Stopwatch
Large magnifying glass
Cardboard
300 ml Beaker
Sunglasses

SAFETY PROCEDURES
USING A MAGNIFYING GLASS TO CONCENTRATE THE SUN'S ENERGY CAN GENERATE VERY HIGH TEMPERATURES. WHEN PERFORMING THIS EXPERIMENT, MAKE SURE THAT STUDENTS DO NOT FOCUS THE SUN'S ENERGY ON ANY PART OF THEIR BODY OR CLOTHING, AS SEVERE BURNS MAY RESULT. ALSO, HAVING STUDENTS WEAR SUNGLASSES WHILE PERFORMING THIS LAB OUTSIDE MAKES IT EASIER TO FOCUS THE SUN'S IMAGE WITHOUT HURTING THE EYES. ALSO, REMIND STUDENTS THAT THEY SHOULD NEVER STARE AT THE SUN WITH THEIR EYEYS, EVEN WITH SUNGLASSES, AS EYE DAMAGE WILL OCCUR.

Procedure

Part 1: High-Temperature Solar and Water

1. Fill a beaker with water, and go outside with a partner to a location where you have a clear, unobstructed view of the Sun.

2. Place your thermometer on the ground, preferably on a section of asphalt blacktop.

3. Put on your sunglasses and then orient yourself so the Sun is at your back.

4. Kneel down near the thermometer, and with help from your instructor, use your magnifying glass to focus the sunlight onto the bulb of the thermometer. The idea is to make the smallest, most concentrated dot of sunlight with the magnifying glass. This can be done by moving the magnifying glass in or out from the bulb of the thermometer, and changing its angle slightly to create the most concentrated dot of energy.

5. As soon as the Sun's focused energy hits the bulb of the thermometer, have your partner begin timing with the stopwatch.

6. Keep the Sun's focused energy on the bulb, and carefully monitor the temperature.

7. As soon as the temperature of the thermometer reaches 100°C (212°F), stop timing. This is the boiling point for water, and the temperature at which steam can be generated to power an electric turbine.

8. Once the thermometer reaches 100°C (212°F), stop heating it with the magnifying glass. CAUTION—CONTINUING TO APPLY THE SUN'S FOCUSED ENERGY ONTO THE THERMOMETER AFTER IT REACHES 100°C (212°F) MAY CAUSE THE THERMOMETER TO BURST.

9. Record the time it took to reach 100°C (212°F) in Table 8-1. Gather three other times from three other groups, and record in Table 8-1. Calculate the average time for all 4 observations.

10. Next, pour a small amount of water onto a portion of blacktop asphalt to make a small puddle. Do not use all of the water in your beaker, as you will need it for Part 2.

11. Use your magnifying glass to focus the Sun onto the puddle, and observe how long it takes for the water to begin to steam.

TABLE 8-1 High Temperature Solar	
Group Names	**Time to Reach Boiling Point of Water 100°C (212°F)**
Average	

Part 2: Solar Furnace

1. Next, place your piece of cardboard down on the asphalt.

2. Use the magnifying glass to focus the Sun's energy onto the cardboard, until it begins to smolder. The flashpoint of paper is 450°F (232°C).

3. Once the cardboard has ignited, use the remaining water in your beaker to extinguish it.

Conclusions

1. What was the average time it took for the temperature of the thermometer to reach the boiling point of water?

2. Explain how Part 1 of the experiment illustrates that solar energy can be used to heat water.

3. What are some problems that might be associated with using high-temperature solar?

4. What design considerations would you have to consider when building a high-temperature solar power station?

5. Did the water poured onto the asphalt begin to steam when concentrated sunlight was focused on it? Do you think you would get the same results if you tried the experiment on a light-colored sidewalk? Why or why not?

6. Was the magnifying glass able to create enough heat to ignite the cardboard?

7. What are some of the ways you think that high-temperature solar can be applied besides heating water to create steam and electricity?

LAB 9
Photovoltaic Cells

Purpose

The purpose of this lab is to observe how photovoltaic cells are used to produce electricity from the Sun. Students will also investigate how the Sun's angle and intensity affects the output of energy from a solar cell.

Materials

Solar cell
Electric meter
Lamp
Protractor
100-Watt light bulb
25-Watt light bulb
2 Coated electric wires with alligator clips
Ring stand with clamp

SAFETY PROCEDURES
HAVE STUDENTS WEAR SAFETY GLASSES OR GOGGLES WHILE PERFORMING THIS EXPERIMENT TO PROTECT THEIR EYES IN CASE THE LIGHT FALLS AND THE LIGHT BULB BREAKS. ALSO, ALLOW TIME FOR THE LIGHT BULBS TO COOL DOWN BEFORE CHANGING THEM. THEY CAN GET EXTREMELY HOT.

Procedure

Part 1: Solar Cell Output

1. Set up your lamp so it is clamped to the ring stand, and is pointing straight down about 24 inches above the tabletop.

2. Put the 25-watt light bulb into the lamp.

3. With the lamp turned off, place your solar cell directly underneath the lamp.

4. Hook up your solar cell to the electric meter using the wires and alligator clips. Make sure you have the positive lead of the solar cell attached to the positive wire (red), and the negative lead of the solar cell attached to the negative wire (black).

5. Attach both red and black wires from the solar cell to the electric meter.

6. Set the electric meter to record Direct Current (DC) amperage.

7. Slide the solar cell directly below the lamp.

8. The angle between the light and the surface of the solar cell in this position is 90 degrees.

9. Turn on the light, and use the electric meter to determine the amperage and voltage output of the solar cell with the 25-watt bulb and record it in Table 9-1.

10. Turn off the light, and allow the bulb to cool down for a few minutes.

11. Once the 25-watt bulb has cooled, replace it with the 100-watt bulb.

12. Turn on the lamp, and determine the amperage and voltage output from the solar cell using the electric meter. Record your data in Table 9-1.

13. Turn off the light. Transfer the data on wattage and voltage for the 100-watt bulb in the row for the 90 degree angle in Table 9-2.

TABLE 9-1 Intensity vs. Solar Cell Output			
Bulb Watts	**Amps**	**Volts**	**Solar Cell Watts**
25			
100			

TABLE 9-2 Angle vs. Solar Cell Output			
Angle	**Amps**	**Volts**	**Solar Cell Watts**
90°			
45°			
0°			

14. Use your protractor to adjust the solar cell so that it is tilted up approximately 45 degrees from the table.

15. While you are holding the solar cell at 45 degrees, have your partner turn on the light, and record the wattage and voltage output in Table 9-2.

16. Next, hold the solar cell so that it is tilted 90 degrees up from the table and record the wattage and voltage output in Table 9-2.

17. Turn off the light. Now calculate the solar cell wattage for the two different light bulbs by multiplying amperage by wattage, and record in Table 9-1.

18. Complete Table 9-2 by calculating the solar cell wattage for each angle.

Part 2: Solar Graphs

1. Figure 9-1 shows the power curve for a solar array over a 3-day period. Use it to answer the following questions.

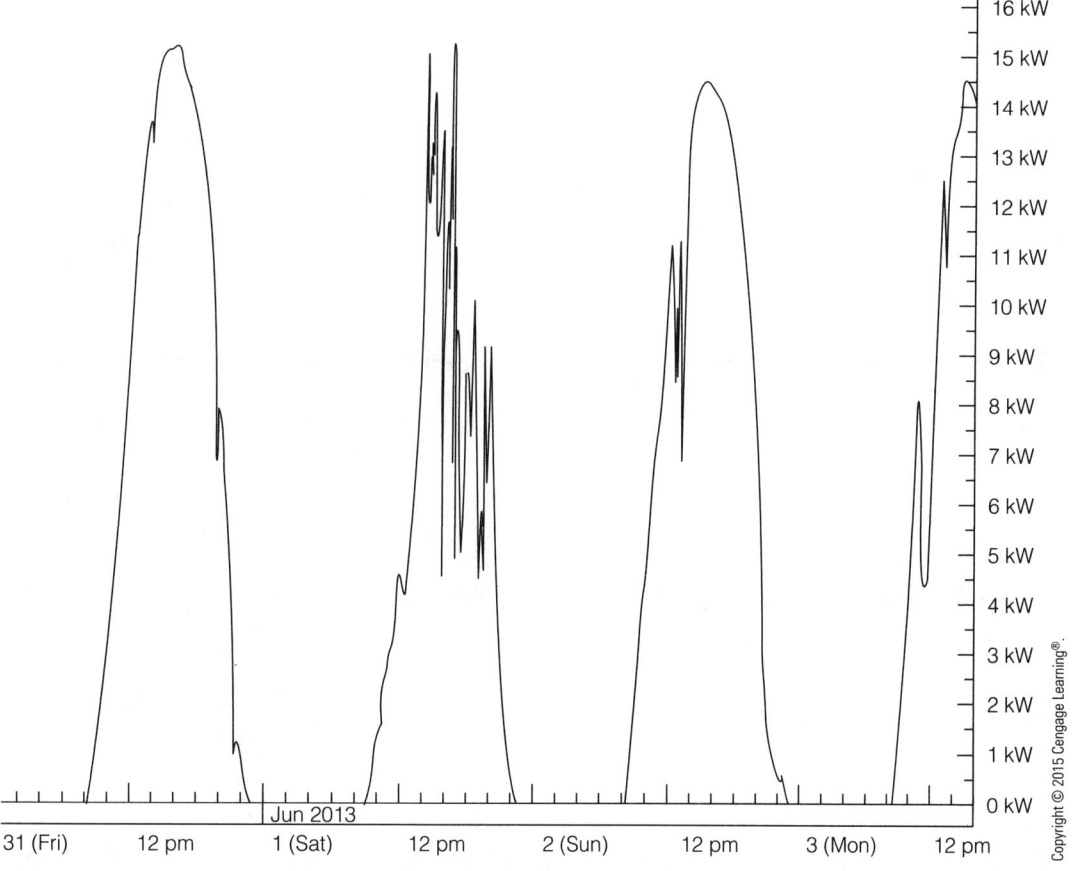

FIGURE 9-1   Solar Array Output

2. What are the units used for solar cell power production?

3. Which day do you think was the clearest, sunniest day? How do you know?

4. Which day was the cloudiest day? How do you know?

5. Do you think any of the three days were cloudless? Why or why not?

6. What do you think happened during the first part of the day on June 2?

7. Between which times of day was the power output of the solar cells the greatest? Why?

Conclusions

1. Which light bulb produced the highest amount of wattage?

2. Which light bulb would represent a cloudy day?

3. How would cloudy days affect the output of energy from a solar cell?

4. Which angle between the surface of the solar cell and the light source produced the highest wattage?

5. Because the angle of insolation changes seasonally, which season would produce the greatest amount of solar power? The least?

6. If one square meter of ground receives an average of 300 watts per square meter of sunlight, how much power in watts could a one-square meter solar panel produce, assuming it is 15% efficient?

7. If the average American family consumes 25 kilowatts per day, use your answer for question 5 to determine how many square meters of solar panels they would require to meet all their electricity needs.

LAB 10
Wind Turbines and Power Curves

Purpose

The purpose of this lab is to construct a scale model of a wind turbine in order to identify its main components. You will also create and analyze a wind turbine power curve to learn how wind velocity affects electricity production.

Materials

Scissors
X-Acto blade
2 ml Disposable plastic pipette
Tape
Toothpick
Graph paper or spreadsheet software
Colored pencils

Procedure

Part 1: Wind Turbine Model

1. Carefully use scissors to cut out all the parts of your wind turbine model shown in Figure 10-1.

2. Cut out the holes and slits in the tower carefully using an X-Acto blade.

3. Assemble the wind turbine tower and secure it using tape.

4. Assemble the nacelle and secure it using tape.

5. Insert the nacelle into the tower with the arrow end sticking out the back.

6. Next, cut out the wind turbine pinwheel in Figure 10-2.

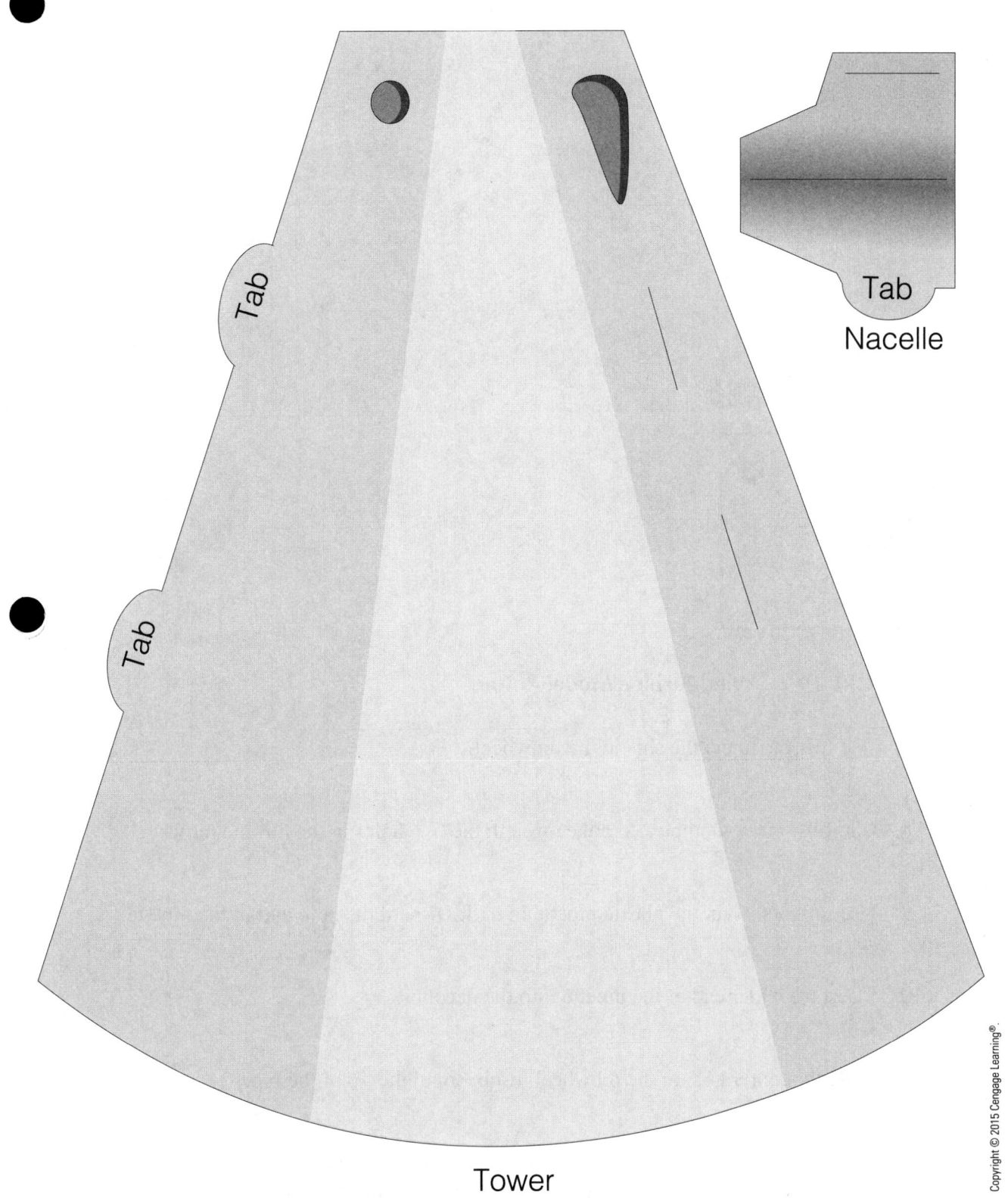

Tab

Nacelle

Tower

FIGURE 10-1 **Wind Turbine Model**

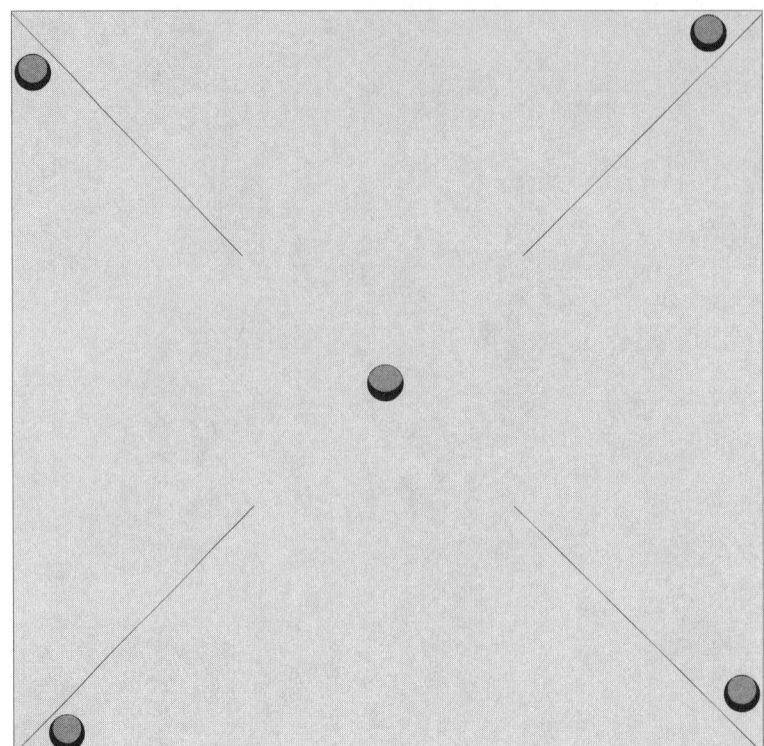

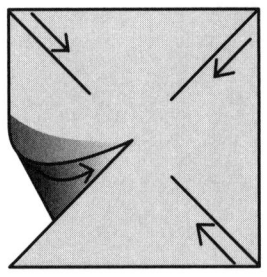

FIGURE 10-2   Wind Turbine Model Rotor

7. Use scissors to cut the slits in the pinwheel.

8. Carefully use a toothpick to poke through the five holes in the turbine pinwheel.

9. Use scissors to cut the plastic pipette in half. Discard the bulb end of the pipette.

10. Insert the wider end of the pipette into the nacelle.

11. Slide the center hole of the pinwheel turbine onto the tip of the pipette.

12. Carefully fold the blades of the pinwheel over and slide the holes near the tips of the blade onto the pipette tip to make the pinwheel. You might have to spin the pinwheel around to loosen it up so it spins freely when you blow on it.

Part 2: Parts of a Wind Turbine

Fill in the missing labels for all the parts of a wind turbine in Figure 10-3, using the following terms: power cables, gear box, transformer, generator, wind, transmission lines, rotor blade, nacelle, and tower.

Part 3: Wind Power Curve

Use the data form Table 10-1 to create a dual line graph showing the relationship between the power output and wind velocity for a wind turbine.

Part 4: National Wind Potential

1. The National Wind Map in Figure 10-4 shows the wind potential for the United States as measured from about 260 feet off the ground. This would be an optimum height for a large scale wind turbine. The average wind speeds are classified into 3 main groups which include Excellent (greater than 9.5 m/s or 31 mph), Great (7–9 m/s or 23–30 mph), and Good (5.5–7 m/s or 18–22 mph).

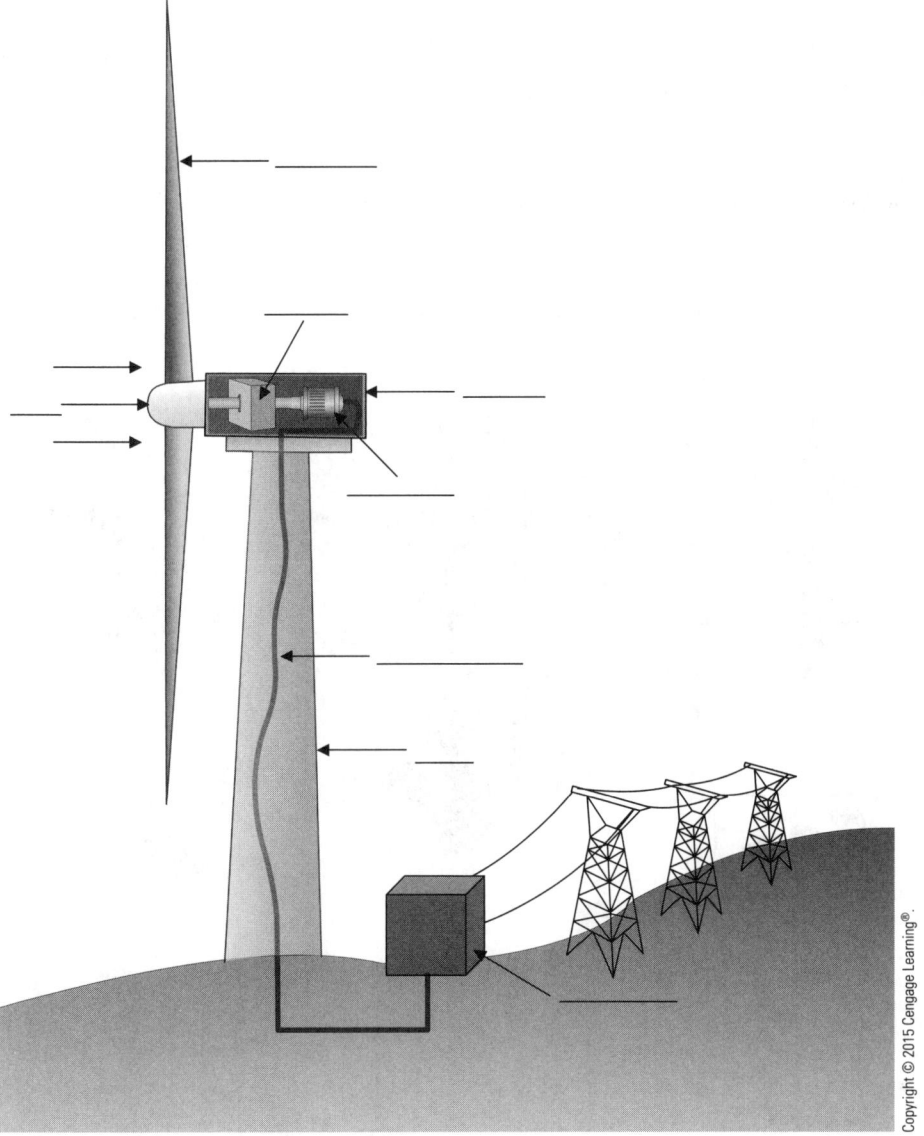

FIGURE 10–3

TABLE 10-1 Power Output and Wind Velocity		
Wind Velocity (miles per hour)	Turbine A Power Output (watts)	Turbine B Power Output (watts)
0	0	0
5	12	3
10	425	110
12	1,200	300
15	1,500	375
20	2,000	500
25	1,800	450
28	1,650	410
33	1,600	400
35	1,500	375
40	1,200	300
45	1,300	325

2. Use 3 different colored pencils to color in the three wind classification zones. Also use your colored pencils to create a key.

3. Use your map to answer conclusion questions.

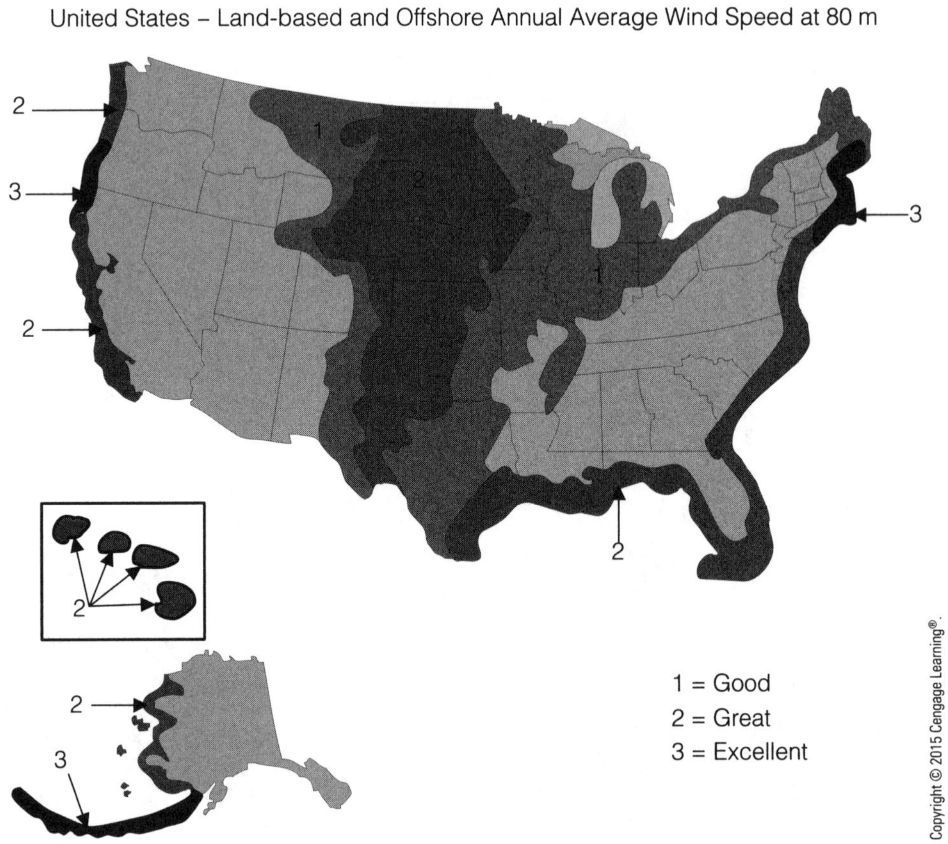
United States – Land-based and Offshore Annual Average Wind Speed at 80 m
2
3
2
1
2
1
3
2
2
2
3
1 = Good
2 = Great
3 = Excellent

FIGURE 10-4

Conclusions

1. Which parts of a wind turbine are housed within the nacelle?

2. What is the general relationship between the length of a wind turbine's rotor blade and the turbine's power output?

3. What are the cut in speeds for the turbines in Part 3?

4. What are the maximum power outputs for both turbines A and B, and at what wind speeds do they occur?

5. How many kilowatts of power does turbine A create with a wind velocity of 15 miles per hour?

6. What happens to the power output of turbine A after the wind speed exceeds 20 miles per hour?

7. The furling speed of a wind turbine is the speed at which the turbine begins to slow down in order to not become damaged by high velocity winds. What is the furling speed of turbine B?

8. Why does the power output of a wind turbine dip, and then go back up after it reaches its furling speed?

9. Which wind turbine in Part 3 do you think has longer rotor blades? Why?

10. Does your state have good wind-power potential? If so, what category does it fall in within the map?

11. Which regions of the United States have the greatest wind-power potential?

12. Which states have excellent wind-power potential?

13. Do you think there is potential for the use of offshore wind turbines in the United States? Why or why not?

LAB 11
Geothermal Heat Exchange

Purpose

The purpose of this lab is to observe how geothermal energy can be used to both cool and heat a building. In this lab, you will construct a model of a simple geothermal heating and cooling system, and observe how heat energy is transferred to and from the ground.

Materials

1 2-liter Plastic soda bottle (with top cut off just below the top shoulder)
1 8-feet Length of ¼ inch flexible plastic tubing
1 Small plastic funnel
2 1,000-ml Beakers
Thermometer
Ring stand and clamp
Ice cubes

Procedure

Part 1: Geothermal Heating

1. Insert the plastic funnel tightly into one end of the plastic tubing.

2. Set the ring stand clamp so it is 24 inches above the base of the ring stand.

3. Insert the funnel into the ring stand clamp and tighten so it holds the funnel in place.

4. Coil the plastic tubing around the inside of the plastic soda bottle, and leave enough at the end to hang over the side (see Figure 11-1).

5. Put the end of the plastic tube into an empty 1,000 ml beaker. This will act as the collection beaker.

6. Fill the other 1,000 ml beaker with 1,000 ml room-temperature water.

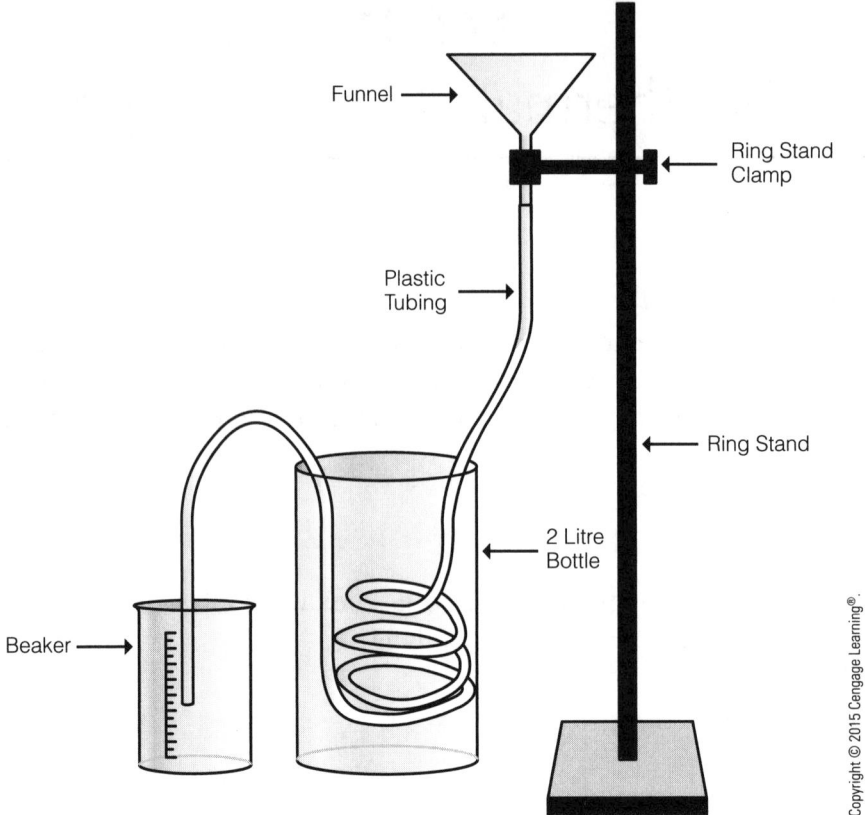

Funnel ⟶

Ring Stand
Clamp

Plastic
Tubing ⟶

Ring Stand

2 Litre
Bottle

Beaker ⟶

FIGURE 11-1

TABLE 11-1 Seasonal Water Temperatures		
Season	**Temperature of Water Before**	**Temperature of Water After**
Winter		
Summer		

7. Insert the thermometer into the room-temperature water, and wait about a minute for the temperature to adjust; then record the temperature of the water for the summer season in Table 11-1.

8. Carefully fill the plastic soda bottle with hot water.

9. Next, pour the room-temperature water from the beaker down into the funnel. Keep pouring until all the water has passed through the tube and has emptied into the other beaker.

10. Record the temperature of the water in the beaker in Table 11-1.

11. Carefully remove the plastic tubing from the soda bottle and empty the hot water out into the sink.

Part 2: Geothermal Cooling

1. Reset the geothermal system, but this time fill the plastic soda bottle with ice cubes. Once the container is full of ice, pour in a small amount of cold water into the plastic soda bottle to make an ice slurry. Make sure the plastic tube coil is completely covered with ice water.

2. Empty out the 1,000 ml collection beaker, and place the end of the tube into the beaker.

3. Refill the other beaker with 1,000 ml room-temperature water and record its temperature in Table 11-1 for the winter season.

4. Carefully pour the water into the funnel until all the water has passed through the system.

5. Record the temperature of the water in the collection beaker.

Conclusions

1. Which part of the experiment acted as the ground?

2. Did the temperature of the water change after it passed through the geothermal system in Part 1? If so, how?

3. Which type of energy transfer method occurred between the water in the tubing and the "ground"?

4. Determine the amount of energy transferred from the water to the ground in Part 1 using the following formula: (end temp − start temp) × (4.18 joules/ml × 1,000 ml).

5. Did the temperature of the water change after it passed through the geothermal system in Part 2? If so, how?

6. How could the transfer of heat energy from ground to the tubing be improved?

7. Determine the amount of energy transferred from the water to the ground in Part 2 using the following formula: (start temp − end temp) × (4.18 joules/ml × 1,000 ml).

LAB 12
Electricity Production and Use in the United States

Purpose

The purpose of this lab is to determine the main sources of energy used to produce electricity in the United States. You will also determine how much energy is lost during electricity production and its transmission.

Materials

Calculator
Colored pencils
Paper or spreadsheet software

Procedure

Part 1: Electricity Flow

1. Using Figure 12-1, fill in the correct data into Table 12-1 for the amount of quads of energy used to produce the nation's electricity. One quad is equivalent to 1 quadrillion BTU's (1×10^{15}).

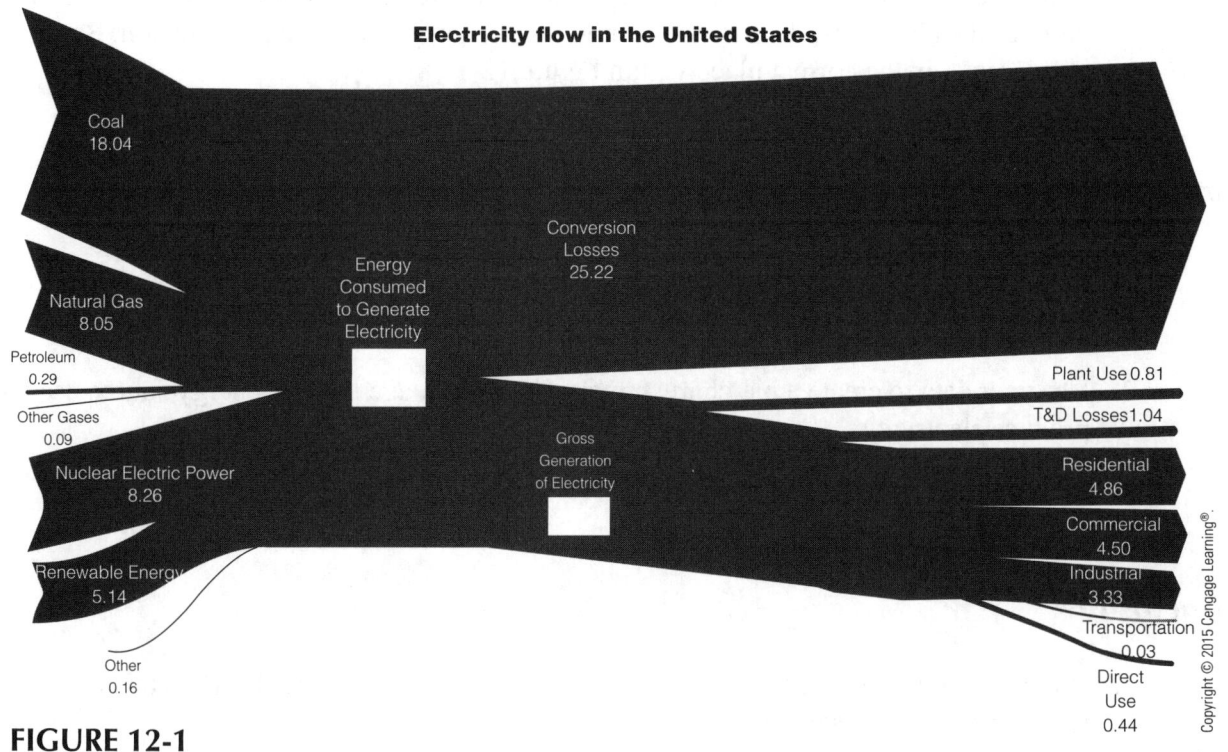

FIGURE 12-1

53

TABLE 12-1 Energy Usage		
Energy Source	Quads	Kilowatt Hours
Coal		
Natural gas		
Petroleum		
Other gases		
Nuclear		
Renewable		
Other		
Total		

2. Convert the data in Table 12-1 into kilowatt hours, knowing that 1 quad $= 2.9 \times 10^{11}$ kilowatt hours.

3. Using the data in Table 12-1, calculate the total amount of energy consumed to produce the electricity used in the United States in quads. Record the answer in Table 12-1 and also in the correct location within Figure 12-1.

4. Use the data in Figure 12-1 to determine the gross amount of electricity consumed in quads by sector (residential, commercial, industrial, etc.). Record your answer in the correct location within Figure 12-1.

5. Using the answer you calculated in step 4, and the total amount of energy consumed to produce the electricity, determine the amount of energy lost during conversion. Record your answer in the correct place within Figure 12-1.

Part 2: Electric Power Pie Chart

1. Convert the data on the amount of quads for each energy source which is used to produce electricity into percentages.

2. Use your data to create a pie chart showing the percentage of each energy source used to produce electricity.

Conclusions

1. Which energy source in the United States produces the most electricity? Is this a renewable energy resource?

2. Are there any environmental problems associated with the energy source you identified in question 1? If so, what are a few examples?

3. What percentage of the total energy resources used to produce electricity are considered fossil fuels?

4. What percentage of total amount of electricity consumed is lost during transmission and distribution (T&D losses)?

5. Determine the efficiency of our electrical production system by dividing the total amount of energy consumed to produce electricity, by the gross generation. Is this a highly efficient system? Why or why not?

LAB 13

Electric Circuits and the Production of Electricity

Purpose

The purpose of this lab is for you to construct simple electric circuits and to be able to identify their main components. You will also observe the differences between parallel and series electric circuits and how they effect electric currents. And you will learn the similarities and differences between electric generators and electric motors.

Materials (per group)

2 Hand-operated generators
Voltage meter
Flashlight light bulb with fixture
1 Black electric wire with alligator clips on both ends
2 Red electric wires with alligator clips on both ends
Simple electric switch

Procedure

Part 1: Basic Electric Circuit

1. All basic circuits consist of a source of electricity, wires to conduct the flow of the electric current, a switch to control the current, and a load.

2. Attach one red wire to the positive wire of the hand generator.

3. Attach the other end of the red wire to one of the terminals on the switch.

4. Attach one end of another red wire to the other terminal of the switch and the other end of the wire to the light bulb fixture.

5. Finally, attach one end of a black wire to the light bulb fixture and the other end to the negative wire of the hand generator.

6. Make sure the switch is in the off position.

7. Begin turning the hand generator slowly, and then slide the switch to the on position.

8. Reverse the direction of how you turn the generator with the switch still on.

9. Now turn the generator very fast.

10. Proceed to Part 2.

Part 2: Parallel Circuit

1. Assemble a simple parallel circuit using the diagram shown in Figure 13-1.

2. Both positive wires from each generator should be clipped to the positive (red) wire of the voltage meter.

3. Both negative wires from each generator should be clipped to the negative (black) wire of the voltage meter.

4. Set the voltage meter to record low DC voltage.

5. First, steadily turn only one generator in a clockwise direction, and then reverse it counterclockwise and observe how the direction of turning the generator effects the voltage.

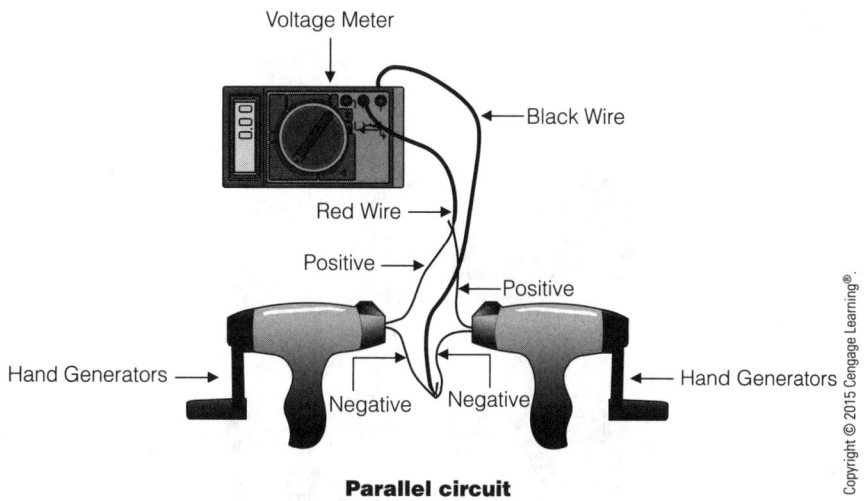

FIGURE 13-1

57

TABLE 13-1 Parallel Circuit Voltage	
Number of Generators	**Volts**
One	
Two	

6. Turn the generator at a slow, steady rate in the correct direction to produce a positive voltage, and observe the voltage it produces on the voltage meter.

7. Next, turn the generator rapidly at a steady state and record its voltage in Table 13-1.

8. Now you and a partner turn both generators at the same time in the same slow, steady rate that you used in Step 6. Record the voltage in Table 13-1.

Part 3: Series Circuit

1. Assemble a simple series circuit using the diagram shown in Figure 13-2.

2. Attach the positive wire from one hand generator to the positive (red) wire of the voltage meter.

3. Attach the negative wire of the hand generator to the positive wire of another hand generator.

4. Finally, attach the other negative wire of the hand generator to the negative (black) wire of the voltage meter.

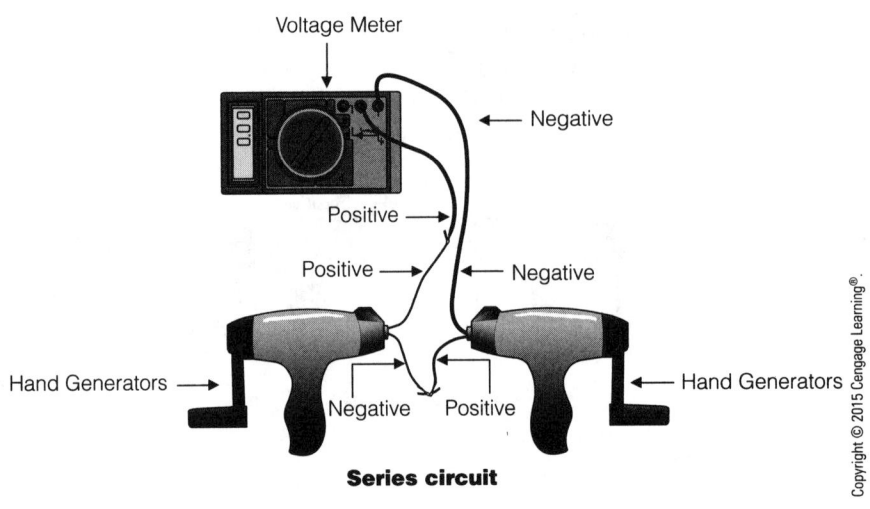

FIGURE 13-2

58

TABLE 13-2 Series Circuit Voltage	
Number of Generators	**Volts**
One	
Two	

5. Turn only one generator at a rapid, steady rate and record its voltage in Table 13-2.

6. Next, you and your partner turn both generators at the same time, at a rapid steady rate and record the voltage in Table 13-2.

Part 4: Electric Generators and Motors

1. Connect the positive lead of one generator directly to the positive lead of the other generator.

2. Connect the negative lead of one generator directly to the negative lead of the other generator.

3. Turn only one generator, and observe what happens.

4. Now disconnect both generators, and attach the positive leads of both generators to the light bulb fixture.

5. Attach both negative leads of both generators to the light bulb fixture.

6. Turn only one generator, and observe what happens

Conclusions

1. What are the four basic parts of an electric circuit?

2. Draw a diagram of a simple circuit.

3. What happened to the light bulb when you reversed the direction of turning the hand generator? Why?

4. What happened to the light bulb when you increased the rate of turning the generator?

5. How did combining two generators effect the voltage for the parallel circuit in Part 2?

6. What happened to the voltage when two generators were turned for the series circuit in Part 3?

7. Multiplying voltage by the amperage of a circuit determines the amount of wattage a load requires. If a typical household electric current consists of 120 volts, how much amperage is required to power a 60-watt light bulb?

8. What happened when you hooked both generators together and turned one of them?

9. What was the other generator acting as?

10. How did connecting the light bulb between the two generators affect how difficult it was to turn the generator to light the light bulb?

LAB 14
Radiation

Purpose

The purpose of this lab is for you to observe how surface properties affect the ability for substances to absorb and re-radiate heat. These basic concepts are important when trying to understand how passive solar energy can be used as an energy source.

Materials (per group)

Black can with foam lid
Silver can with foam lid
Heat lamp
Ring stand
2 Thermometers
Stopwatch
Graph paper or spreadsheet software
Colored pencils

SAFETY PROCEDURES
HAVE STUDENTS WEAR SAFETY GLASSES OR GOGGLES WHILE PERFORMING THIS EXPERIMENT TO PROTECT THEIR EYES JUST IN CASE THE LIGHT FALLS AND THE LIGHT BULB BREAKS. ALSO, INSTRUCT STUDENTS TO BE CAREFUL WHEN TURNING THE LIGHT OFF OR HANDLING THE LIGHT FIXTURE BECAUSE IT CAN GET VERY HOT.

Procedure

1. Set both cans so they are approximately 24 inches away from the heat lamp.

2. Make sure the heat lamp is off and is on the same level as the cans.

3. Adjust the light so both cans are receiving equal amounts of light radiation (see Figure 14-1).

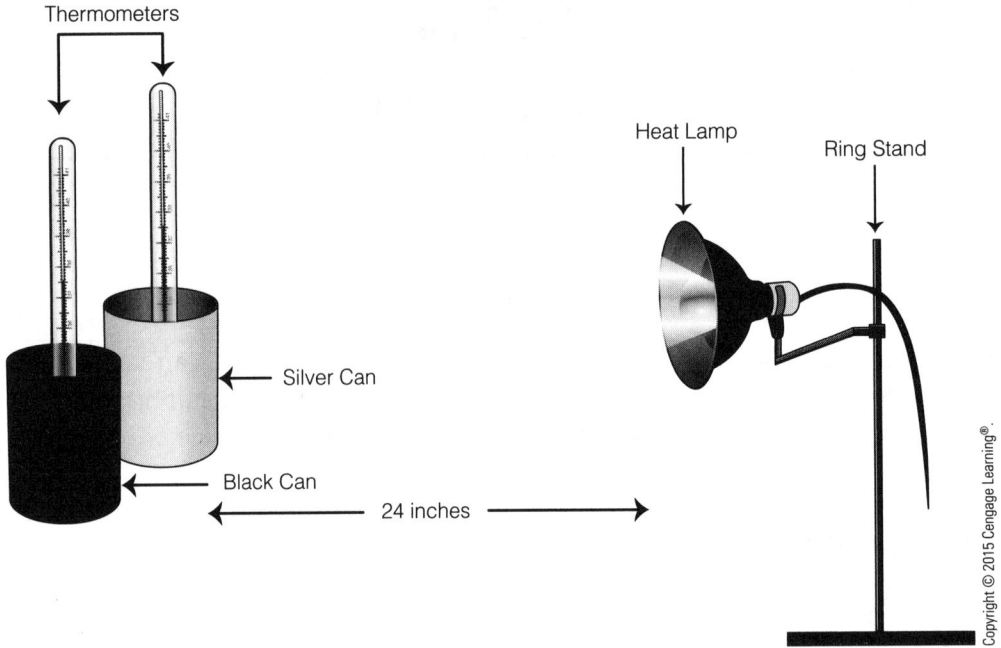

Thermometers

Silver Can

Black Can

← 24 inches →

Heat Lamp

Ring Stand

Copyright © 2015 Cengage Learning®.

FIGURE 14-1

4. Insert a thermometer into each of the cans, making sure that the bulbs of the thermometers do not touch the sides of the cans.

5. Wait for the temperature for each can to stabilize, and then record these as the start temperatures in Table 14-1.

6. Have your instructor check your experimental set-up, and once given the OK, turn on your light and begin timing with your stopwatch.

7. Record the temperature of each can in Table 14-1 every minute for 10 minutes.

8. After 10 minutes, turn off your light, and carefully point it away from the cans.

9. Continue to record the temperature of each can for another 10 minutes.

10. Once your experiment is complete, clean up your workstation according to your instructor's directions.

11. Use the data in Table 14-1 to create a dual line graph showing the temperature for each can over the 20 minute experiment.

TABLE 14-1 Can Temperatures		
Time (Minutes)	Black Can (Temp.)	Silver Can (Temp.)
0		
1		
2		
3		
4		
5		
6		
7		
8		
9		
10		
11		
12		
13		
14		
15		
16		
17		
18		
19		
20		

Conclusions

1. Calculate the rate of heating for the black can during the first 10 minutes of the experiment using the following formula: change in temp/10 minutes.

2. Calculate the rate of heating for the silver can during the first 10 minutes of the experiment.

3. Which can absorbed energy at a greater rate?

4. Calculate the rate of heating for the black can during the last 10 minutes of the experiment.

5. Calculate the rate of heating for the silver can during the last 10 minutes of the experiment.

6. Which can re-radiated energy at a higher rate?

7. What change occurred within the can to indicate that it was absorbing energy?

8. What change occurred within each can to indicate that they were re-radiating energy?

9. What is a good color for a substance to have in order for it to absorb and re-radiate energy at a high rate?

10. How do the wavelengths of energy absorbed by a substance compare to the wavelengths of energy re-radiated?

LAB 15
Exothermic Reaction

Purpose

The purpose of this lab is for students to observe a simple exothermic reaction. An exothermic reaction produces heat and is an important part of understanding thermochemistry and the laws of thermodynamics.

Materials (per group)

Sodium hydroxide (lye)
300 ml Beaker
Thermometer
pH test strips
Safety glasses
Plastic gloves
Balance or scale

SAFETY PROCEDURES
STUDENTS MUST WHERE SAFETY GLASSES AND PLASTIC GLOVES WHILE PERFORMING THIS EXPERIMENT. THE USE OF SODIUM HYDROXIDE (LYE) NEEDS TO BE PERFORMED WITH CAUTION AS IT CAN BURN THE SKIN IF TOUCHED. PLEASE FOLLOW ALL SAFETY GUIDELINES AS OUTLINED BY YOUR INSTRUCTOR.

Procedure

1. Before beginning this lab, please make sure you are wearing your safety glasses and protective gloves.

2. Add 100 ml of water to a beaker.

3. Carefully weigh out 2.0 grams of sodium hydroxide.

4. Place a thermometer into the beaker of water and record the temperature of the water in Table 15-1.

5. Use a pH test strip to determine the pH of the water and record in Table 15-1.

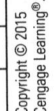

TABLE 15-1 Temperature and pH			
Sart Temp (°C) T₁	End Temp (°C) T₂	Start pH	End pH

6. Carefully add the 2.0 grams of sodium hydroxide into the beaker of water.

7. Gently stir the sodium hydroxide with the thermometer while observing its temperature.

8. Record the highest temperature of the solution in Table 15-1.

9. Use another pH test strip to determine the pH of the solution after the reaction and record in Table 15-1.

10. Clean up your workstation as directed by your instructor.

Conclusions

1. What is an exothermic reaction?

2. Where did the energy come from to produce the heat in this experiment?

3. Determine the heat of reaction in joules produced by the experiment using the following formula:

mass of water (100 grams) × change in temperature (T_2 × T_1) × 4.18 joules/g × °C = heat of reaction

4. Did the pH of the water change after the addition of the sodium hydroxide? If so, how?

LAB 16
A Simple Battery

Purpose

The purpose of this lab is for you to construct a simple battery and identify its main parts. You will also utilize different electrolyte solutions and measure the different voltage outputs they produce. Understanding how electrochemical cells produce electric currents and how they can be improved is an important aspect of modern renewable energy research.

Materials (per group)

5 Dimes
5 Pennies
Vinegar
Table salt (NaCl)
Paper towels
Small plate
Voltage meter
200 ml Beaker
Scale or balance
Scissors
Ruler
2 ml Plastic disposable pipette
Graph paper or spreadsheet software

Procedure

Part 1

1. Use the scissors to carefully cut out 10 2-cm $\times$ 2-cm squares from a paper towel.

2. Place the paper towel squares on the plate near one side and then use the pipette to moisten them with tap water.

3. Place one penny down on the plate near the center, and cover it with one of the moist pieces of paper towel.

4. Place a dime on top of the first penny, and then cover it with another damp piece of paper towel.

5. Continue to make your battery stack by placing another penny on top of a dime, making sure to place a moist paper towel between each coin. The last coin at the top should be a dime.

6. Your simple battery is now ready to be tested.

7. Set your voltage meter to low DC voltage, and place the tip of one lead of the meter on the penny at the bottom of the stack, and the other lead to the dime at the top.

8. Record the voltage that the battery is producing in Table 16-1. If the voltage is recording as a negative value, reverse the leads so that the one touching the dime is touching the penny and vice versa.

9. Take apart your battery, dispose of the ten pieces of paper towels, and wipe off your plate so that it is dry.

10. Pour 50 ml of vinegar into a 200 ml beaker.

11. Cut out 10 more squares the same size as you did in step 1.

12. Place the squares on the plate near one side.

13. Use a pipette to moisten the squares with vinegar from the beaker.

14. Build another battery by stacking dimes on top of pennies, with pieces of paper towels moistened with vinegar. Make sure you end with a dime on top of the battery.

15. Use the voltmeter to determine the voltage output of the battery, and record it in Table 16-1.

TABLE 16-1 Battery Voltage	
Electrolyte	Volts
Tap water	
Vinegar	
Vinegar/Salt	

16. Take apart your battery, dispose of the ten pieces of paper towels, and wipe off your plate so that it is dry.

17. Weigh out 8 grams of table salt (NaCl) using a scale or balance.

18. Add the table salt to the beaker containing the vinegar, and stir it until it is dissolved.

19. Cut out 10 more paper towel squares, and lay them on the plate near one side.

20. Moisten the squares with the vinegar/salt solution, and build another battery.

21. Use the voltmeter to determine the voltage produced by the battery, and record it in Table 16-1.

22. Clean up your workstation as directed by your instructor.

Part 2

Use your data from Table 16-1 to create a bar graph that shows the relationship between electrolyte and voltage for your battery.

Conclusions

1. What basic parts of an electrochemical cell did the coins act as in your simple battery?

2. What role did the tap water, vinegar, and vinegar/salt solution play in your battery?

3. If modern pennies are made of mostly zinc and modern dimes are coated with nickel, which acted as the anode and which as the cathode?

4. Which coin lost electrons (were oxidized)?

5. Which coin gained electrons (were reduced)?

6. Which way did the electrons flow in your battery (from which coin to which coin)?

7. Did changing the electrolyte solution affect the voltage output for the battery?

8. Which electrolyte produced the highest amount of voltage?

LAB 17
Photosynthesis and Respiration

Purpose

In this lab, you will observe how plants undergo both photosynthesis and respiration, and will also identify the conditions in which these processes occur.

Materials

Healthy Elodea plants
Clean test tubes with corks or caps (four for each group)
Test tube racks
Tape for labels
Aluminum foil
Drinking straws
Glass jars or 500-ml beakers
Artificial light or window
pH indicator solution*

SAFETY PROCEDURES
WHEN PREPARING OR WORKING WITH THE PH INDICATOR SOLUTION, ALWAYS WEAR SAFETY GLASSES!

Procedure

1. Each group should pour two jars or beakers one-half full with pH indicator solution. Charge one jar with carbon dioxide by blowing gently into the solution, using a straw as demonstrated by your instructor. When the solution is charged with enough carbon dioxide, it will turn to a yellow-green color. Label this jar with the tape "Carbon Dioxide Added (Yellow)." Label the other jar "No Carbon Dioxide (Blue)."

---

*Before the laboratory is conducted, a stock solution of bromothymol blue must be made by mixing 0.5 grams of bromothymol blue powder with 500 ml of water. Once this is mixed, add 9 drops of sodium hydroxide (NaOH). Next, mix 20 ml of stock solution with 480 ml of water, and add 5 drops sodium hydroxide (NaOH). This is the solution that will be used to carry out the experiment.

2. Next, use the tape to label four test tubes A, B, C, and D, and place in a test tube rack.

3. Place a four- or five-inch section of healthy Elodea plants in test tubes A and C.

4. Carefully pour the Carbon Dioxide Added (Yellow) solution into test tubes A and B. Make sure to completely fill the test tubes and cap them.

5. Carefully pour the No Carbon Dioxide (Blue) solution into test tubes C and D. Make sure to completely fill the test tubes and then cap them.

6. Use aluminum foil to completely wrap test tubes C and D so as not to allow any light to enter.

7. Place all test tubes in a rack and put them in an area exposed to natural light.

8. Fill in Table 17-1 by placing a check mark for all of the correct parameters for each test tube, and let them sit undisturbed overnight.

9. After one day, observe the conditions in all four test tubes, and record the end color for each test tube on Table 17-1.

10. After you have recorded your results, clean up your experiment by following the specific clean-up procedures outlined by your instructor.

TABLE 17-1 End Color Chart							
	Light	**No Light**	**CO₂**	**No CO₂**	**Plant**	**No Plant**	**Start Color**	**End Color**
A								
B								
C								
D								

Conclusions

1. What occurred in the pH indicator solution to cause it to turn from blue to yellow when you blew into it through the straw?

2. Write the chemical reactions for both photosynthesis and respiration.

3. Which test tubes acted as controls in your experiment and why?

4. Which test tube underwent photosynthesis? Explain how your results proved this.

5. Which test tube underwent respiration? Explain how your results proved this.

6. Describe the specific conditions that cause plants to undergo photosynthesis and respiration.

LAB 18
Primary Productivity: How Much Does a Field Weigh?

Purpose

The purpose of this lab is to have you determine the productivity of a 1-acre agricultural field by calculating the amount of biomass it produces. You will then utilize your data to estimate the amount of biomass that the field contains.

Materials

Paper bags
String, 36-inches long
Balance or scale

Procedure

Before you begin this activity, try and estimate how much all of the biomass in a 1-acre field weighs. Have your instructor take down each student groups' guesses before beginning this lab exploration.

1. Go to the designated area outside of your school with your instructor to collect your biomass sample. You will be determining the productivity of a managed grassland ecosystem, so you need to collect a sample of grass. At the area designated by your instructor, select a section of healthy growing grass.

2. Surround the grass you have selected with your piece of string that has been shaped into a square roughly 6-inches long on each side.

3. Carefully remove all of the grass, including the root system, from within your string square. Shake off as much soil from the roots as possible and place your biomass sample in a paper bag. Make sure to label your bag with your group's name.

4. Now return to your classroom with your biomass sample. Because biomass is the total dry weight of plant material, you must let your sample dry out overnight. Place your bag, with the top open, in a warm sunny location in your classroom in order to dry it out completely.

TABLE 18-1 Biomass Weights			
Sample #	Mass in Grams	Mass in Pounds	Calculations
Average			

5. After your biomass sample has been completely dried out, carefully determine its mass to the nearest tenth of a gram. Record your answer on Table 18-1. Get the data from two other groups in your class so you have the weight of at least three samples. Record these measurements on Table 18-1.

6. Convert all three masses of your samples into pounds by dividing your mass in grams by 453.6. Record your answers on Table 18-1.

7. Calculate the average for all three samples and record your answer on Table 18-1.

Conclusions

1. Define the term biomass.

2. Using the data from Table 18-1, calculate how much biomass on average, 1 square foot of managed grassland weighs in pounds. Remember that your sample represents 36 square inches (6" × 6"), and 1 square foot equals 144 square inches (12" × 12"). Record your answer below. Show your work.

3. Next, you must calculate the area in square feet of a 1-acre field, knowing that a 1-acre square is 208.7 feet by 208.7 feet. Record your answer below. Show your work.

4. Determine the weight of the football field by multiplying the weight of 1 square foot of your biomass by the answer you calculated in Question 3. Record your answer below. Show your work.

5. To determine how close your "guess" was to the actual weight you calculated in Question 4, you must use the percent deviation calculation. The percent deviation of a measurement is determined by the difference between the actual value and your guess, divided by the actual value, then multiplied by 100. A percent deviation of 3 percent or less is considered accurate. Determine the percent deviation of your "guess" below, and show all of your work.

6. Briefly describe the process used to determine the productivity of an agricultural field growing grass.

LAB 19
Biomass Gasification

Purpose

The purpose of this lab is to observe how any form of plant biomass can be used to produce a combustible gas known as syngas. This process is known as *biomass gasification* and it occurs as a result of pyrolysis, which can be used to produce a renewable source of both heat and power.

Materials

Wood pellets, dried chopped grass, or woodchips
Bunsen burner or alcohol burner
Vent hood
Ring stand with test tube clamp
Pyrex test tube
Rubber stopper with hole
Glass tubing (approximately 2 inches in length)
Matches
Safety glasses

SAFETY PROCEDURES
THIS LAB WILL NEED TO BE PERFORMED IN A VENTILATION HOOD TO ALLOW FOR THE SAFE REMOVAL OF FUMES PRODUCED DURING THE EXPERIMENT. ALSO, THE USE OF SAFETY GLASSES BY EVERY STUDENT DURING THE EXPERIMENT IS REQUIRED. BECAUSE STUDENTS ARE WORKING WITH AN OPEN FLAME, MAKE SURE THEY KNOW THE PROPER SAFETY PROCEDURES FOR LIGHTING AND EXTINGUISHING A BUNSEN BURNER OR ALCOHOL BURNER. ALSO, MAKE SURE STUDENTS WITH LONG HAIR HAVE IT TIED BACK TO AVOID CONTACT WITH AN OPEN FLAME. IF YOU ARE USING MATCHES, MAKE SURE TO DIP THEM IN WATER OR RUN THE MATCH HEAD UNDER WATER BEFORE DISPOSING OF THEM.

Procedure

Part 1

1. Fill your test tube about ¼ full with some dried biomass which can be in the form of wood pellets, dried chopped grass, or woodchips.

2. Carefully insert the glass tubing into the hole of the rubber stopper, assuring a tight fit. Leave about 1 to 1½ inches sticking out from the stopper.

3. Insert the rubber stopper tightly into the test tube.

4. Attach the test tube to the clamp and ring stand. Adjust the test tube so that it sits at a 45 degree angle.

5. Carefully light the burner.

6. Adjust the height of the ring stand so that the base of the test tube is just above the flame (see Figure 19-1).

7. When the biomass is first heated, water vapor and smoke will be produced and vent out through the glass tube. After a few minutes, all the air inside the test tube will be driven out, and pyrolysis will begin.

8. Use a match to try and ignite the vapor coming out of the glass tube. It may take a few tries, but eventually it will light. Once the vapor is combustible, syngas is being produced.

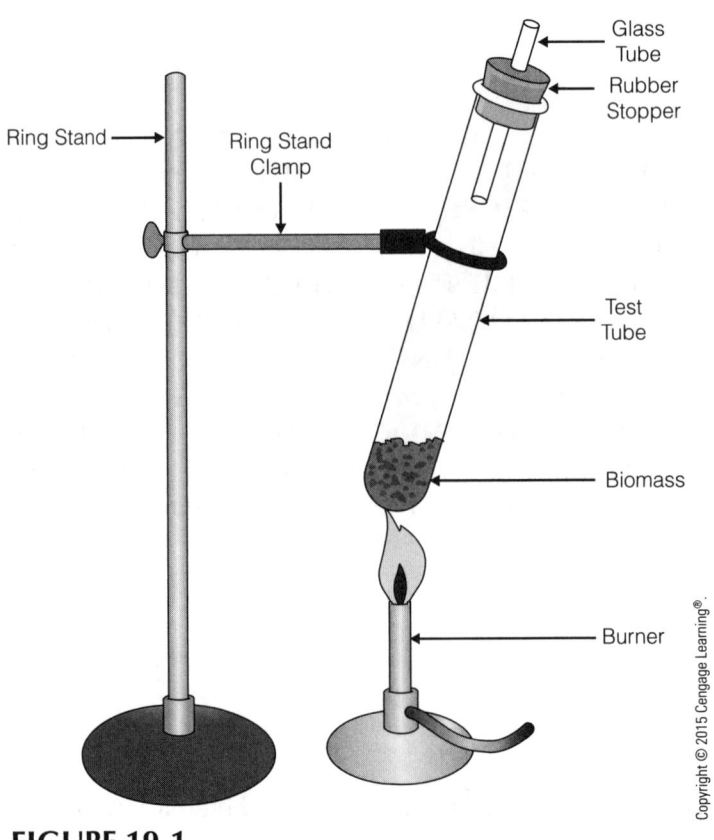

FIGURE 19-1

TABLE 19-1   Energy Data
Switchgrass = 7,000 BTUs per pound and yields 5 tons per acre
Shrub Willow = 8,000 BTUs per pound and yields 5 tons per acre
#2 Fuel Oil = 140,000 BTUs per gallon
School's Annual Fuel Use = 100,000 gallons per year

9. Carefully turn off the burner, and observe the contents of the test tube. Wait until the apparatus is completely cooled before disassembling it per your teacher's instructions.

Part 2

1. Use the data in Table 19-1 to answer the following questions.

2. How much switchgrass would you need to produce the same amount of heat the school uses from burning #2 fuel oil?

3. How many acres of switchgrass would you need to grow to be able to heat the school for one year?

4. How much shrub willow would you need to produce the same amount of heat the school uses in one year?

5. How many acres of shrub willow would you need to grow to supply the school with heat for one year? (Remember it takes three years to grow one crop.)

Conclusions

1. Did the contents of the test tube change after pyrolysis? If so, how?

2. How does pyrolysis differ from combustion?

3. What are three products produced by biomass gasification?

4. Besides being flammable, what is another danger of syngas?

5. The process of biomass gasification is 97% efficient. If 2,000 pounds of biomass is gasified each day to produce 100 kilowatts of heat and power, how many pounds of biochar is left over?

LAB 20
Bioenergy Crops

Purpose

The purpose of this lab is for students to grow different types of crops used for producing energy. They will also observe the effects of soil salinization on plant growth and calculate the percent germination for bioenergy crop seeds, which is an important part of improving the efficiency of agricultural production.

Materials (per group)

2 Petri dishes
Paper towels
Scissors
23 Bioenergy crop seeds (switchgrass, soybeans, sunflowers, mustard, reed canary grass, gama grass, indian grass, bluestem, wheatgrass)
Table salt (NaCl)
200 ml Beaker
2 Disposable pipettes
Scale or balance
Potting soil
Small plastic pot
Masking tape
Sunny window or plant growth light

Procedure

Part 1: Bioenergy Crop

1. Fill a plastic pot with potting soil.

2. Poke three holes in the soil about 1-inch deep.

3. Collect 3 seeds from your instructor of the species of bioenergy crop you are growing.

4. Place 1 seed in each hole, and cover with soil.

5. Carefully water your pot to just moisten the soil.

6. Place your pot under the grow light or near the window.

7. Use masking tape to label the pot with your group name.

Part 2: Percent Germination and Soil Salinization

1. Use a lid of 1 of the 3 petri dishes to trace out 4 circles on a piece of paper towel.

2. Cut out the 4 circles with scissors, and place 1 circle of paper towel in the bottom of the 2 petri dishes.

3. Add 100 ml of tap water to a 200 ml beaker.

4. Weigh out 0.5 grams of table salt (NaCl) and add to the 100 ml of water.

5. Stir with a pipette until the salt is dissolved. This will make a 0.5% salt solution.

6. Use the masking tape to label the lid of each petri dish. Label one dish "Tap Water" and the other dish "0.5% Salt."

7. Retrieve 20 seeds of the bioenergy crop you are testing from your instructor, and place 10 seeds onto each paper towel in both petri dishes. Make sure to space them apart from one another. Record the number of seeds planted in each dish in Table 20-1.

8. Carefully cover over the seeds with the other pieces of paper towel you cut out.

9. Use the pipette that you mixed the salt solution with, to add a small amount of water to the 0.5% petri dish; just enough to soak the paper towels. Do not flood with water!

TABLE 20-1 Germination Data			
Treatment	Number of Seeds Planted	Number of Seeds Germinated	Percent Germination
Tap Water			
0.5% Salt Solution			

10. Use another pipette to add tap water to the other petri dish.

11. Cover both dishes, and place them in the area designated by your instructor.

12. Observe the seeds over the next week, and when your instructor tells you, record how many seeds germinated out of the 10 in each dish in Table 20-1.

13. Clean up your experiment per your instructor's directions.

Conclusions

1. Which type of bioenergy crop did you plant?

2. Is your bioenergy crop used to produce oil or for biomass?

3. Is your bioenergy crop also a food crop?

4. Is your bioenergy crop a perennial or annual plant? Why would this make a difference to the farmer?

5. Which experimental treatment for procedure B acted as the control?

6. Was there a difference in the percent germination between the 2 treatments? If so, which one was lower?

7. What can you conclude about the effect on soil salinization on the rate of germination for plants?

LAB 21
Ethanol Biofuel Fermentation and Distillation

Purpose

The purpose of this lab is to have students use the processes of fermentation and fractional distillation to convert simple sugars into ethanol biofuel.

Materials (per group)

250 ml Erlenmeyer flask
125 ml Erlenmeyer flask
Rubber stopper with one hole
Rubber stopper with two holes
Plastic tubing ~12 inches in length
250 or 125 ml Beaker
Yeast
Sugar
Stirring rod
Distilled water
Incubator
Pipette
10 ml Graduated cylinder
Balance or scale
Thermometer
Distillation condenser
Safety glasses
Bromothymol blue (0.04% aqueous solution)
Pasteur salts (optional) – *the addition of Pasteur's Salts helps to buffer acids that are produced as a result of the fermentation process. To prepare a solution of Pasteur's salts add the following to 860 ml of distilled water: 10.0 g of ammonium tartrate, 2.0 g of potassium phosphate, 0.2 g calcium phosphate, 0.2 magnesium phosphate.*

Procedure

Part 1: Fermentation

1. Add 175 ml of distilled water into the 250 ml Erlenmeyer flask.

2. Add 20.0 grams of sugar into the flask, and use the stirring rod to mix it in the water until it is totally dissolved.

3. Add 2.0 grams of dried baker's yeast into the water-sugar solution, and use the stirring rod to evenly mix it.

4. Fill a beaker with the bromothymol blue solution. Bromothymol blue is a solution that turns from blue to yellow/green when exposed to an acid.

5. Place the stopper in the Erlenmeyer flask.

6. Insert the tubing into the stopper, and make sure it is an airtight seal.

7. Put the other end of the tubing in the bromothymol blue as shown in Figure 21-1.

8. Place the whole apparatus in an incubator set to about 30°C. As the yeast begins to ferment the sugar, it will produce carbon dioxide gas as a byproduct. As the gas exits the flask and out through the tubing, it will turn the bromothymol blue solution yellow/green. The fermentation process should take about 3–4 days to be complete, and when all the sugar has been consumed by the yeast, the bromothymol blue solution will begin to turn back to blue.

9. Once the fermentation process is complete, the yeast should have settled to the bottom of the flask. Carefully disassemble the apparatus, and save the yeast/ethanol solution for Part 2.

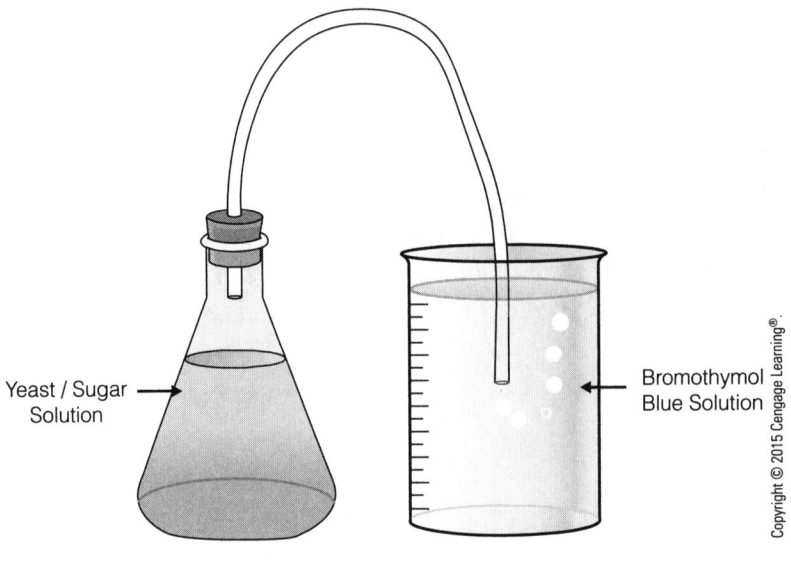

Yeast / Sugar Solution

Bromothymol Blue Solution

Copyright © 2015 Cengage Learning®.

FIGURE 21-1

Part 2: Ethanol Distillation

1. Determine the mass of a dry, 10 ml graduated cylinder.

2. Use a pipette to remove 10 ml of the yeast/ethanol solution and place it in the graduated cylinder. Make sure to avoid drawing up any sediment that has settled at the bottom of the flask.

3. Determine the mass of the 10 ml sample of yeast/ethanol solution.

4. Subtract the dry weight of the empty graduated cylinder from the total weight of the solution and the graduated cylinder. Record your results in Table 21-1.

5. Carefully decant the yeast/ethanol solution into a 125 ml Erlenmeyer flask without disturbing the settled yeast.

6. Insert a thermometer into the hole of the rubber stopper, and a small section of tubing into the other hole. You may wish to seal off the opening where the tubing enters the condenser to prevent loss of vapor to the air. This can be done with aluminum foil or plastic wrap.

7. Insert the stopper into the 125 ml flask.

8. Assemble the distillation condenser apparatus as shown in Figure 21-2.

9. Set the hot plate temperature to heat the ethanol solution to between 78° and 88°C. Make sure not to exceed 88°C.

10. Distill the ethanol until you have collected at least 10 ml.

11. Determine the density of 10 ml of ethanol distillate

TABLE 21-1 Ethanol Density Data			
Sample	Mass (grams)	Volume (cm^3)	Density (g/cm^3)
Yeast/Ethanol Solution			
Ethanol Distillate			

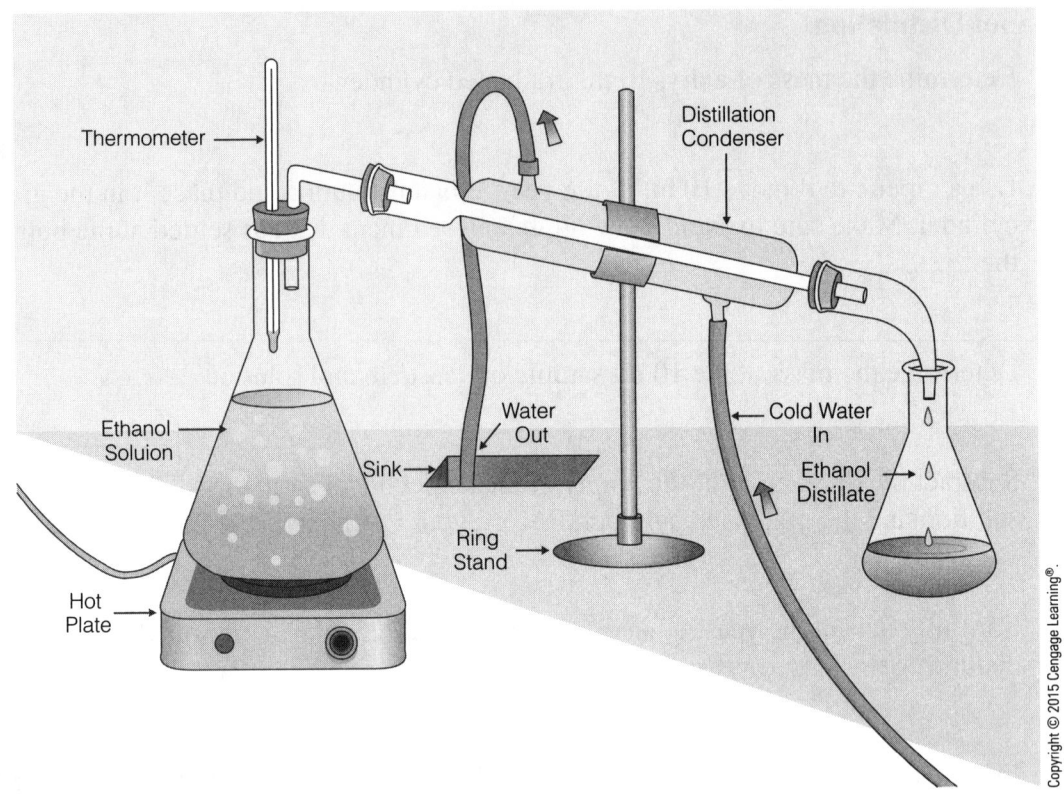

FIGURE 21-2   Distillation Apparatus

Conclusions

1. Write a balanced equation for the fermentation of one molecule of glucose into carbon dioxide and ethanol.

2. Describe two purposes of the bromothymol blue used in this lab.

3. Determine the percent deviation of your ethanol distillate density, considering the average density of nearly pure ethanol is 0.8 g/cm^3.

LAB 22
Cellulosic Ethanol

Purpose

The purpose of this lab is to have students identify the most effective pretreatment methods for extracting glucose from cellulosic biomass. Improving the process of breaking down cellulose into simple sugars is an important step in producing cellulosic ethanol, an important biofuel.

Materials (per group)

Dried switchgrass or other bioenergy crop biomass
Scissors
8–50 ml Plastic culture tubes with lids
Test tube rack
300 ml Beaker
Hot plate
Amylase enzyme solution (6 grams of amylase powder dissolved in 100 ml of water)
1% Sulfuric acid solution
Scale or balance
16 Glucose urine test strips
1 ml Disposable pipette
Masking tape

SAFETY PROCEDURES
PLEASE USE CAUTION WHEN USING THE HOTPLATE. BE CAREFUL WHEN USING THE HOTPLATE TO BOIL WATER.

Procedure

1. Fill the 300 ml beaker about half full with tap water and place on the hotplate.

2. Turn the hotplate on full, and bring the water in the beaker to boiling.

3. Number your test tubes 1 through 8.

4. Use the scissors to cut up your biomass into small pieces.

5. Place your tubes in the test tube rack, and add 20 ml of water to each tube.

6. Using a scale to weigh each portion, place 1 gram of biomass in tubes 1 through 5 and in tube 8. Make sure the grass is completely immersed under the water in the tubes. Tubes 6 and 7 should have no grass in them.

7. Carefully place tubes 1 and 4 in the beaker of boiling water for 10 minutes.

8. After 10 minutes of boiling, turn off the hot plate, and allow tubes 1 and 4 to cool.

9. Use a pipette to add 0.5 ml of the amylase enzyme solution to tubes 1, 2, 7, and 8.

10. Use another pipette to add 0.5 ml of 1% sulfuric acid solution to tubes 4, 5, 6, and 8.

11. Place the lids on all tubes tightly, and gently mix them. Make sure the grass is totally immersed under the water in the tubes.

12. Use the glucose test strips to test the glucose levels in each of the 8 tubes, and record the value in Table 22-1, under "Glucose Level Day 1."

13. After 24 hours, test the 8 tubes again using the glucose test strips, and record the value in Table 22-1, under "Glucose Level Day 2."

14. Clean up your experiment according to your instructor's directions.

TABLE 22-1 Biomass Glucose Yield			
Tube Number	**Treatment**	**Glucose Level Day 1**	**Glucose Level Day 2**
1	Grass – Heat, Enzyme		
2	Grass – No Heat, Enzyme		
3	Grass Only		
4	Grass – Heat, Acid		
5	Grass – No Heat, Acid		
6	Acid Only		
7	Enzyme Only		
8	Grass, Enzyme, Acid		

Conclusions

1. Which treatment produced the highest glucose level on day 2?

2. Which test tubes acted as controls?

3. How do you think cellulose is broken down naturally?

4. What other pre-treatments do you think you might be able to use to increase glucose production from biomass?

5. Do you think this experiment showed the potential for using any biomass to produce sugars for the fermentation of biofuels?

6. What is the cellulose polymer composed of?

7. Which produced the highest amount of glucose, the enzyme or the acid?

8. Where is the amylase enzyme found within the human body? What function does it perform?

9. What other organisms' enzymes might you want to experiment with to use for biofuel production?

LAB 23
Energy Content of Biofuels Lab

Purpose

The purpose of this lab is to have students determine the energy content of 3 liquid biofuels, identify which has the highest energy density, and compare it to a fossilfuel-based transportation fuel.

Materials (per group)

1 Small 8 oz. soda can
Graduated cylinder
12-inch Section of string
Paperclip
Ring stand
Ring stand test tube clamp
Ethanol in an alcohol burner
Butanol in an alcohol burner
Methanol in an alcohol burner
Lamp oil in an alcohol burner
Thermometer
Safety glasses
Stopwatch
Matches
Balance or scale

SAFETY PROCEDURES
PLEASE BE CAREFUL WHEN WORKING AROUND THE OPEN FLAME DURING THIS EXPERIMENT. MAKE SURE IF YOU HAVE LONG HAIR THAT IT IS TIED BACK, AND BEWARE OF LOOSE CLOTHING TOUCHING THE FLAME. ALSO REVIEW THE CORRECT PROCEDURE FOR BOTH LIGHTING AND EXTINGUISHING YOUR BURNER. IF YOU ARE USING MATCHES, MAKE SURE TO DIP THEM IN WATER OR RUN THE MATCH HEAD UNDER WATER BEFORE DISPOSING OF THEM.

Procedure

Part 1

1. Tie the section of string to the paperclip, and affix it to the ring stand test tube clamp, with the paperclip hanging from the string.

2. Hang the soda can from the paper clip.

3. Place an alcohol burner under the soda can as illustrated in Figure 23-1.

4. Add 100 ml of room-temperature water into the soda can.

5. Light the burner and quickly perform a flame test with the ethanol, and adjust the height of the soda can by lowering or raising the test tube clamp so the tip of the flame just touches the center of the bottom of the soda can.

6. Safely extinguish the burner.

7. Weigh the ethanol burner, and record its start mass in Table 23-1.

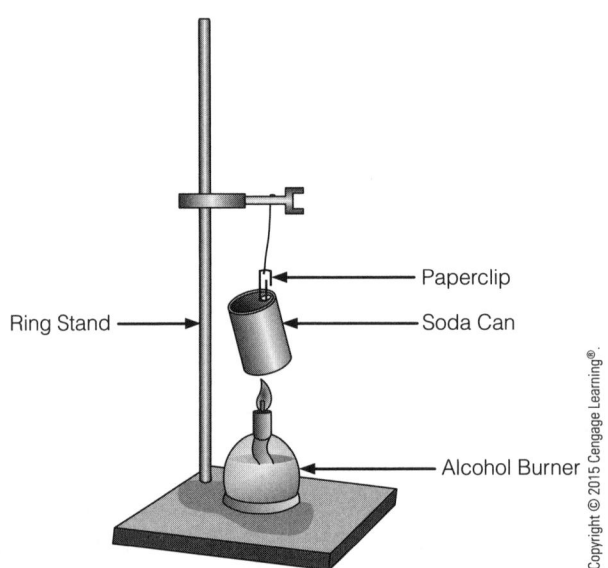

FIGURE 23-1   Heat Content Apparatus

TABLE 23-1 Heat Content Data						
Fuel	Start Temperature Degrees C (T_1)	End Temperature Degrees C (T_2)	Change in Temperature Degrees C $(T_2 - T_1 = \Delta T)$	Start Mass of Fuel in grams (M_1)	End Mass of Fuel in grams (M_2)	Mass of Fuel Used in grams $(M_1 - M_2 = \Delta M)$
Ethanol						
Methanol						
Butanol						
Lamp Oil						

8. Use the thermometer to determine the temperature of the water in the can and record it in Table 23-1.

9. Place the ethanol burner under the can, and carefully light it and heat the water in the can for 5 minutes.

10. After 5 minutes, extinguish the flame and record the temperature of the water.

11. Next, weigh the mass of the ethanol burner and record it in Table 23-1.

12. Repeat the same procedure using the methanol, butanol, and lamp oil.

Part 2

1. Next you will use the data you collected in Table 23-1 to determine the energy content of each fuel in joules per gram. Because you used 100 ml of water, which weighs 100 grams, and it takes 4.187 joules to raise 1 gram of water 1°C, then the amount of joules given off by the fuel can be expressed by the following calculation:

$$\text{Energy released in joules} = \Delta T \, (°C) \times 100 \text{ grams} \times 4.187 \text{ joules/gram/°C}$$

2. Record the results of the energy released by each fuel in Table 23-2.

3. Enter the change in mass for each fuel (ΔM) into Table 23-2.

4. Calculate the energy content for each fuel by using the following formula:

$$\text{Energy content (joules/gram)} = \text{energy released (joules)/mass of fuel burned (grams)}$$

TABLE 23-2 Energy Content of Fuels			
Fuel	Energy Released in joules	Mass of Fuel Burned (grams)	Energy Content in joules/gram
Ethanol			
Methanol			
Butanol			
Lamp Oil			

Conclusions

1. Write below the chemical equations for ethanol, methanol, and butanol.

2. Which biofuel did you determine to have the highest energy content? Which had the lowest?

3. Which biofuel would you recommend to be used to replace gasoline? Why?

4. What potential errors could have altered the outcome of this experiment?

5. As a result of this experiment, do you believe that biofuels have the potential to replace gasoline? Why or why not?

LAB 24
Vegetable Oil Extraction

Purpose

The purpose of this lab is to have students perform the cold method of oil extraction used to produce vegetable oils for the production of biodiesel.

Materials (per group)

Unsalted peanuts or soybean seeds
Mortar and pestle, food processor, or blender
250 ml Beaker
Stirring rod
Methanol
Balance or scale
Funnel
Filter paper
100 ml Glass graduated cylinder
Safety glasses

Procedure A: Density of Methanol

1. First you will determine the density of methanol.

2. To do this, determine the mass of a dry 100 ml graduated cylinder and record it in Table 24-1 under "Methanol."

3. Carefully add 20 ml of methanol to the graduated cylinder, and record the volume in Table 24-1.

TABLE 24-1 Density Data					
	Mass of Empty Cylinder (grams)	Mass of Cylinder and Liquid (grams)	Mass of Liquid (grams)	Volume of Liquid (ml)	Density (g/cm³)
Methanol					
Methanol/Oil Solution					

4. Place the graduated cylinder with the 20 ml of methanol on the scale and record its mass in Table 24-1.

5. Use your data to calculate the density of methanol to the nearest 100th of a gram/cm³ and record in Table 24-1.

Procedure B: Oil Extraction

1. Use a scale or balance to weigh out 75 grams of unsalted peanuts or soybean seeds.

2. Use the mortar and pestle, food processor, or blender to chop or crush the seeds into very small pieces. The more they are chopped up, the better your results will be.

3. Pour the chopped seeds or nuts into 250 ml beaker.

4. Add 100 ml of methanol to the beaker with the chopped seeds or nuts.

5. Carefully stir the seeds and methanol for about 1 minute.

6. Place your beaker in the area designated by your instructor, and let it sit for 24 hours.

Procedure C: Oil Analysis

1. Fold your filter paper per your instructor's directions, and place in the funnel.

2. Place the funnel in the top of a clean, dry 100 ml graduated cylinder.

3. Determine its mass and record in Table 24-1 under Methanol/Oil Solution.

4. Carefully pour the methanol/oil solution from your beaker of seeds and nuts into the funnel without spilling any seeds or debris into the filter.

5. Let the solution settle.

6. Next, determine the mass and volume of the methanol/oil solution, and record in Table 24-1.

7. Use the information from Table 24-1 to derive the density of the methanol/oil solution.

8. Clean up your experiment according to your instructor's directions.

Conclusions

1. Did the appearance of the methanol change after 24 hours? If so how?

2. How did the density of the methanol/oil solution compare to the density of the methanol?

3. If you were using this process to extract oil from seeds, how would you separate the methanol from the oil so that the methanol can be reused?

4. The actual density of vegetable oil is about 0.91 g/cm^3 and for methanol it is 0.79 g/cm^3. Using these values calculate the percent deviation for each using the values you determined for each substance:

    Percent deviation = (difference between accepted value and measured value)/
    accepted value $\times$ 100

5. If you let the solution sit long enough, which substance will settle to the bottom of the cylinder?

LAB 25
Biodiesel Production and Physical Properties

Purpose

The purpose of this lab is for students to learn the basic steps used for the production of biodiesel derived from vegetable oil. They will then compare the physical properties of their biodiesel with that of petroleum diesel and vegetable oil.

Materials (per group)

Sodium hydroxide (lye)
Wide range pH test strips
Methanol (100 ml per group)
Vegetable oil (400 ml per group)
Petroleum diesel (200 ml per group)
Sodium hydroxide titration solution (0.5 NaOH dissolved in 500 ml distilled water)
2–10 ml Graduated cylinder
100 ml Graduated cylinder
250 ml Graduated cylinder
2–5 mm Hard plastic bead
50 ml Beaker
300 ml Beaker
1,000 ml Beaker
2–250 ml Erlenmeyer flask
2 ml Disposable pipettes
Safety glasses
Plastic or rubber gloves
Hot plate
Balance or scale
Thermometer
Isopropyl alcohol (10 ml per group)
#7 Rubber stopper
Stopwatch
Ruler
Ice

Procedure

Part 1: Titration Procedure

Titration of the feedstock oil is necessary to determine the correct amount of catalyst to use for the transesterfication process.

1. Before beginning this part of the lab, please make sure you are wearing your safety glasses and protective gloves.

2. Carefully add 10 ml of isopropyl alcohol to a 50 ml beaker.

3. Use a disposable pipette to add 1 ml of the oil that will be used to add the biodiesel to the isopropyl alcohol in the beaker.

4. Use a pH test strip to test the pH of the oil/alcohol solution, and record in Table 25-1.

5. Add 1 ml of the sodium hydroxide titration solution to the isopropyl alcohol in the beaker, and stir.

6. Test the pH of the solution using a new pH test strip, and record its value in Table 25-1. Continue to add 1 ml of oil to the alcohol solution and test its pH until the pH of the solution falls between 8 and 9.

7. Once the pH of your alcohol/oil solution reaches a pH of between 8 and 9, proceed to Part 2.

TABLE 25-1	pH of Oil/Alcohol Solution
ml of NaOH solution	pH

Copyright © 2015 Cengage Learning®.

Part 2: Biodiesel Production

1. Add 200 ml of feedstock oil to a 250 ml Erlenmeyer flask.

2. Place the flask with oil on a hotplate, and insert a thermometer into the oil. Carefully heat it until the temperature of the oil is 120°F (49°C). DO NOT LET THE OIL'S TEMPERATURE GO ABOVE 120°F (49°C)! If the oil gets too hot it can become flammable!

3. Add 200 ml of methanol to a 300 ml beaker.

4. Weigh 2.5 grams + N of sodium hydroxide, where the N represents the amount of ml you determined in Part 1. For example if it took 2 ml to turn the pH of the alcohol solution in Part 1 to fall between 8 and 9, then you would add 4.5 grams of sodium hydroxide (2.5 + 2) to the methanol. Avoid contact of the sodium hydroxide with your skin as it may cause burns.

5. Carefully add the sodium hydroxide to the methanol in the beaker. Use a stirring rod to mix the sodium hydroxide with the methanol until dissolved. This may take 5 to 10 minutes. The resulting solution is called methoxide.

6. Once the methoxide is dissolved, carefully pour the oil, which has been heated to 120°F (49°C), into a 250 ml Erlenmeyer flask.

7. Now add 40 ml of the methoxide solution to the oil.

8. Tightly insert the rubber stopper into the top of the flask, and then vigorously shake the oil/methoxide together for about 2 minutes.

9. Place the flask down and allow it to settle. The result of the transesterfication process which is occurring in the flask will be the production of biodiesel and glycerin. The glycerin will settle as a thin layer of amber-colored liquid at the bottom of the flask.

10. Clean up your workstation as instructed, and save the biodiesel you made for use in Parts 3 and 4. You can also save the glycerin for use in Lab 26, Soap From Biodiesel.

Part 3: The Density and Viscosity of Biodiesel

1. Next you will determine the density of biodiesel and compare it to the density of raw vegetable oil and petroleum diesel.

TABLE 25-2 Density and Viscosity of Fuels						
Fuel	Mass of Empty Cylinder (grams)	Mass of Liquid and Cylinder (grams)	Mass of Liquid (grams)	Volume of Liquid (ml)	Density (g/cm^3)	Relative Viscosity (cm/sec)
Vegetable Oil						
Biodiesel						
Petroleum Diesel						

2. Weigh a dry and empty 250 ml graduated cylinder, and record its mass in Table 25-2.

3. Add 200 ml of vegetable oil to the graduated cylinder.

4. Determine the mass of the oil and cylinder and record in Table 25-2.

5. Next, you will determine the relative viscosity of the vegetable oil by dropping a 5 mm hard plastic bead into the graduated cylinder filled with oil, and time how long it takes to travel to the bottom of the graduated cylinder. Perform this test twice, and record the average settling time in Table 25-2.

6. Use a ruler to determine the height of the oil within the graduated cylinder to the nearest 10th of a centimeter.

7. Calculate the relative viscosity of the vegetable oil by dividing the height of the oil in centimeters by the average settling time in seconds. Record in Table 25-2.

8. Return the vegetable oil to its original container, and then wash and dry the graduated cylinder.

9. Repeat steps 3–8 to determine the density and relative viscosity of the biodiesel you made in Part 2. Record your data in Table 25-2.

10. Return the biodiesel to its original container, and then wash and dry the graduated cylinder.

11. Repeat steps 3–8 to determine the density and relative viscosity of petroleum diesel. BE CAREFUL NOT TO INHALE ANY OF THE FUMES FROM THE DIESEL FUEL WHEN WORKING WITH IT. Record your data in Table 25-2.

12. Clean up your workstation as instructed.

Part 4: Biodiesel Cloud Point

1. Pour 6 ml of biodiesel into a 10 ml graduated cylinder.

2. Pour 6 ml of petroleum diesel into a 10 ml graduated cylinder.

3. Carefully place both graduated cylinders into a 1,000 ml beaker.

4. Slowly add ice to the beaker, to surround the bottom half of each graduated cylinder.

5. Add a small amount of water to the beaker to make an ice slurry surrounding the bottom half of the graduated cylinders within the beaker.

6. Gently insert the bulb of a thermometer into the graduated cylinder of the biodiesel, being careful to not cause it to overflow.

7. Observe the temperature at which the biodiesel begins to become cloudy.

8. Clean up your workstation as instructed.

Conclusions

1. What type of vegetable oil did you use?

2. How many gallons of oil can be produced by the feedstock crop grown to produce the oil that you used to make your biodiesel?

3. How many milliliters of biodiesel did you produce from the 200 ml of vegetable oil?

4. What percentage of your vegetable oil produced glycerin?

5. How did the density of your biodiesel compare to that of vegetable oil and petroleum diesel?

6. How did the viscosity of your biodiesel compare to that of vegetable oil and petroleum diesel?

7. What happened to your biodiesel and the petroleum diesel when you placed them in the ice bath?

8. What was the cloud point temperature of the biodiesel?

9. How do you think the cloud point of biodiesel would impact its use for transportation?

LAB 26
Soap from Biodiesel

Purpose

The purpose of this lab is for students to learn how soap can be made from the byproduct of producing biodiesel. When biodiesel is made from vegetable oil, approximately 10–20% of the original volume of the oil will produce glycerin as a byproduct of the transesterfication process. The glycerin can then be used to make soap.

Materials (per group)

Sodium hydroxide (lye)
Glycerin (you can use the glycerin you produced from making biodiesel in Lab 25)
Essential oils (optional if you want to make scented soap)
200 ml Beaker
300 ml Beaker
100 ml Graduated cylinder
Safety glasses
Plastic or rubber gloves
Hot plate
Balance or scale
Thermometer
Stirring rod
Small plastic yogurt container
Scissors

SAFETY PROCEDURES
STUDENTS MUST WHERE SAFETY GLASSES AND PLASTIC GLOVES WHILE PERFORMING THIS EXPERIMENT. THE USE OF SODIUM HYDROXIDE (LYE) NEEDS TO BE PERFORMED WITH CAUTION AS IT CAN BURN THE SKIN IF TOUCHED. ALSO CAREFUL ATTENTION MUST BE PAID TO THE HEATING OF THE GLYCERIN SOLUTION ON THE HOT PLATE AND TO NOT EXCEED THE REQUIRED TEMPERATURE. ALSO, AVOID INHALING ANY FUMES DURING THIS EXPERIMENT. PLEASE FOLLOW ALL SAFETY GUIDELINES AS OUTLINED BY YOUR INSTRUCTOR.

Procedure

1. Add 25 ml of water to the 200 ml beaker.

2. Carefully weigh 4.0 grams of sodium hydroxide, and add to the water in the beaker. Use a stirring rod to mix the sodium hydroxide within the water until dissolved.

3. Heat the sodium hydroxide/water solution to 100°F (37°C) using a hot plate. Be careful to not inhale any fumes.

4. Add 100 ml of glycerin to a 300 ml beaker.

5. Carefully remove the sodium hydroxide/water solution from the hot plate.

6. Place the beaker of glycerin on the hot plate and heat until it is 100°F (37°C).

7. Pour the sodium hydroxide/water solution to the glycerin. Continue to heat the solution at 100°F (37°C), and stir for about 10–20 minutes. The solution may begin to foam, which is OK. Do not heat the solution past 100°F (37°C).

8. After about 20 minutes, add about 1–2 ml of some essential oil to the mixture to make scented soap. Then carefully pour the glycerin solution into a plastic yogurt container.

9. Allow the soap to cool for about 24 hours, or until it solidifies. You may have to use scissors to cut away the plastic container to release the soap.

Conclusions

1. Describe how the physical properties of the glycerin changed during the experiment.

2. Explain how making soap from biodiesel makes the process of biodiesel production more efficient.

3. How did your soap turn out? Was it like a bar of soap you can buy at the store? Did it work?

4. What do you think you would change in this experiment to make liquid hand soap?

LAB 27
Biogas Generator

Purpose

The purpose of this activity is for you to observe another fermentation process, which is the result of naturally occurring bacteria. The creation of biomethane, a commonly called natural gas, can be produced by bacteria living in anoxic (oxygen-free) conditions. Researchers have utilized this fermentation process to create natural gas in a device called a *biogas generator*. In this activity you will construct a simple biogas generator to observe the action of bacteria, and learn how biogas is produced.

Materials

1 Plastic 12-ounce soda or juice container with lid
Animal manure (cow, sheep, goat, horse, or bagged manure from a garden store)
Large balloon
Rubber band
Corn syrup
1,000 ml Beaker
250 ml Graduated cylinder
Measuring cup
Measuring spoon
Balance or scale
Dechlorinated water (It is important to use water that has no chlorine in it for this experiment to be successful. Spring water or distilled water can be purchased at a grocery store, or tap water can be dechlorinated by using a dechlorinating agent available at an aquarium supply store.)

Procedure

1. Collect all the required materials and bring them to your work area. Mix approximately 175 grams of manure with 175 ml of dechlorinated water (1:1 mixing ratio) in the 1,000 ml beaker.

2. Add 1 tablespoon of corn syrup into the beaker.

3. Carefully pour the mixture into the soda bottle so as to completely fill it. Then cover the opening at the top of the bottle with the balloon. Make sure the balloon is brought all the way down around the opening to assure an airtight fit. You may need to reinforce the seal between the balloon and the bottle, by tightly wrapping a rubber band around the bottleneck and the balloon.

4. Label your container with your group names, and place it in a well-ventilated, dark, warm place designated by your instructor. A greenhouse or outside location works well, so the smell of the manure does not disturb your classroom.

5. You will observe your biogas generator over a period of a week or two. At first, carbon dioxide gas will be produced by aerobic (oxygen-loving) bacteria, this can be released from the balloon by temporarily breaking the seal of the balloon. Eventually, anaerobic (no-oxygen) bacteria, called methanobacteria, will begin to digest the manure and produce methane gas which will begin to collect in the balloon. If the balloon becomes filled, release the gas by lifting the side of the balloon away from the opening of the bottle. The balloon should refill with gas in a few days.

SAFETY PROCEDURES
ALTHOUGH THE METHANE GAS PRODUCED IN THE BALLOON CONTAINS A HIGH WATER CONTENT, METHANE GAS IS FLAMMABLE, SO PLEASE KEEP THE BALLOON AWAY FROM ANY OPEN FLAME OR HEAT SOURCE.

Conclusions

1. Explain why it is necessary to use water that has no chlorine in it for this experiment.

2. Why was it necessary to mix corn syrup in the generator?

3. Explain why carbon dioxide gas was generated first by bacteria before the methane gas was produced.

4. Where did the microorganisms that produced the methane gas come from in this experiment?

5. Describe why biogas is considered to be a renewable, agriculture-based fuel.

6. What are two advantages that biogas generation might have for our society?

7. What are other fuels that might be produced by living organisms?